Cosmic Explorers

COSMIC EXPLORERS

Scientific Remote Viewing, Extraterrestrials, and a Message for Mankind

Courtney Brown

A DUTTON BOOK

DUTTON
Published by the Penguin Group
Penguin Putnam Inc., 375 Hudson Street, New York, New York 10014, U.S.A.
Penguin Books Ltd, 27 Wrights Lane, London W8 5TZ, England
Penguin Books Australia Ltd, Ringwood, Victoria, Australia
Penguin Books Canada Ltd, 10 Alcorn Avenue, Toronto, Ontario, Canada M4V 3B2
Penguin Books (N.Z.) Ltd, 182–190 Wairau Road, Auckland 10, New Zealand

Penguin Books Ltd, Registered Offices:
Harmondsworth, Middlesex, England

First published by Dutton, a member of Penguin Putnam Inc.

First Printing, July, 1999
10 9 8 7 6 5 4 3 2 1

LIBRARY OF CONGRESS CATALOGING-IN-PUBLICATION DATA:
Brown, Courtney
Cosmic explorers : scientific remote viewing, extraterrestrials, and a message for mankind / Courtney Brown.
p. cm.
ISBN 0-525-94430-3
1. Life on other planets. 2. Extraterrestrial anthropology. 3. Remote viewing (Parapsychology) I. Title
QB54.B758 1999 98-53341
001.942—dc21 CIP

Printed in the United States of America
Set in Palatino and Bureau Agency
Designed by Eve L. Kirch

This book is printed on acid-free paper ∞

Contents

PART III. THE WAR

PART IV. THE GREYS AND THE GALACTIC FEDERATION

PART V. THE MARTIANS

PART VI. HUMANITY'S CHOICE OF DESTINY

OPTION #1

OPTION #2

SECTION ONE

PART I

BACKGROUND

Chapter 1

A Turning Point

Humanity is at a turning point in its history. Down through the ages, sages in every civilization have said that there is more to life than merely the flesh and bones within which we dwell. We have souls, and physical death is but a blink in our awareness. This concept is easy enough to entertain. From religious beliefs to reports of life after death, all of us have felt, at one time or another, that there is something beyond our mortal selves.

The continuum of life, though, may be vaster than we ever realized. For some time many scientists have believed that we are not alone in the universe. Scattered throughout the far reaches of space, other life forms exist. Until recently we have been unable to contact these beings. Yet if we had a method that combined the intuitive perception of the sages with the rigor of modern science, then humans could find their true place in the cosmos.

What needs to be examined is a component of the current "scientific" worldview that maintains that only that which is physical is real. The basis of this relatively new religion of scientific atheism is the belief that consciousness is limited to the physical mind, and that when the brain stops functioning, consciousness ceases to exist. When this happens, the personality of the individual is gone forever. But what if the sages from the past are right? The demise of scientific atheism would cause people everywhere to turn inward, in order to seek that which resides beyond. A new scientific age not divorced from the spiritual would dawn.

To accomplish this, a new method, a new tool for exploration, is needed. Based on the explorations of myself and others, I believe such a method exists. This new and surprisingly accurate method of data collection is called "remote viewing." There is now a new scientific field of consciousness that specializes in the study of this phenomenon. People can be trained to use remote viewing to collect information across time and space. Remote viewing is not easy to do, and to use it with consistent accuracy requires extensive training and practice. Explicit procedures have been designed to aid communication between the physical mind and what many call the "subspace" mind (the soul). Souls exist in subspace, that vast domain diversely referred to by mystics and theologians as the etheric realm, heaven, or the afterlife. Humans can learn to become directly aware of their subspace aspects, normally hidden from them until they die and their bodies drop away.

Remote viewing has its origins in the procedures developed largely by Ingo Swann, working under contract at SRI International (formerly Stanford Research Institute) in a program that was funded by various United States governmental agencies, notably the Defense Intelligence Agency (DIA). This program began in the early 1970s and continued until 1989. In 1990 the government transferred its financial support to a program housed at Science Applications International Corporation (SAIC).

The original primary researchers were Russell Targ and Harold Puthoff. Their work in the program involved contact with a number of scientific luminaries, including Charles Tart. Edwin May worked in the program since the mid 1970s, and he became the project director in 1986. When subsequent government funding for remote-viewing research switched to SAIC, Dr. May continued as director.

The current understanding of remote viewing is based on a rich history of experimentation and discovery. In these two decades the literature on remote viewing—both historical and otherwise—has grown to be quite extensive. Over the past few years a number of scientists and remote viewers have risked public ridicule and their professional reputations to pursue research in this new and controversial field.

While this is not the appropriate setting to present a complete history of remote viewing, some recently published books contain detailed historical background of both the scientific and military investigations into psi phenomena generally, and remote viewing in particular. Readers who are interested in this background should read *The Conscious Universe: The Scientific Truth of Psychic Phenomena* by Dean Radin (HarperEdge 1997), and *Remote Viewers: The Secret History of America's Psychic Spies* by Jim Schnabel (Dell 1997). I also strongly recommend interested readers to examine the Spring 1996 volume of the *Journal of Scientific Exploration* (Volume 10, number 1), dedicated almost entirely to the scientific history of remote viewing.

For all its intelligence uses, though, it is remote viewing's ability to penetrate the subspace realm that has led to knowledge of an entirely different sort. Most human efforts to contact other worlds have focused on radio signals or other "hard science" data. Yet what we have been assuming is that extraterrestrials are limited to our own fairly unimpressive modes of communication. What has been discovered through repeated sessions of remote viewing is that the pathway to other worlds lies in the subspace realm.

Remote-viewing evidence suggests that many extraterrestrial species are highly telepathic. Indeed, the human species may be unique in that we have such difficulty perceiving things with our souls while we are physical beings. We already know the universe is a very complex place. In this light, demanding that other life forms communicate only the way we do is tantamount to demanding that the wind blow the way we say.

By using the tools of consciousness, we may be able to fulfill our potential and contact more advanced forms of life. By developing knowledge on the subspace level (that is, the level of the soul), we could protect ourselves by making educated choices regarding our own destiny. Humanity needs vision of all sorts to prosper in a complex universe, and that total vision includes the ability to perceive beyond the physical/subspace divide. In a complex universe, global awareness could not only open new avenues of knowledge, but protect us from harm. If we had the ability to interact with other beings, our human energies could propel us into a future in which we determine our own destiny.

These are uncertain times. As is characteristic of all history, events will happen in our future that will be unexpected. We need to add a degree of control to our future evolution. Control is not obtained through continued ignorance but by increased awareness. Remote viewing is one way we can increase our knowledge of the universe.

THE PLAN OF THIS BOOK

Part I is an overview of the mechanics of Scientific Remote Viewing. It provides the basis for understanding the chapters that follow. Some readers who read my earlier book, *Cosmic Voyage,* will find the section involving types of remote-viewing data familiar. But there are many other elements that have never been published before. Explaining the mechanics helps remove the mystery from the actual sessions that are the heart of this book.

Part II contains a series of remote-viewing sessions using verifiable targets. These chapters are included so that readers can see how the mechanics of Scientific Remote Viewing work with targets about which solid data is known. The targets cover a wide range of substantive areas, and even if the discussion of the extraterrestrials that follows challenges the belief system of some readers, the verifiable targets will give everyone something to think about.

Section 2 of this book contains four parts in which I present new remote-viewing data involving extraterrestrials currently interacting with Earth in varying degrees. The chapters in Section 2 contain information that is crucial for everyone who has an open mind regarding these matters to understand. It contains much of the substantive basis upon which we, as humans, must decide the course of our collective future.

This book is not designed to make people feel comfortable. It is crisis, not confirmation, that assists our species as we make important evolutionary advances. This book is structured to force people to confront ideas that do not conform with pre-existing paradigms.

The truth regarding these issues will not come easily. But the future potential of humankind rests with our success in creating this shift in our collective thought. With a little prodding, humanity may eventually understand and accept these spiritual and scientific ideas, but we do not have the luxury of waiting. As I explain in the pages that follow, our options for the future may soon be dramatically restructured for the worse. Through our own actions we will define and create our destiny.

Chapter 2

SCIENTIFIC REMOTE VIEWING

The method of remote viewing that is the focus of this book began to evolve in earnest in 1996 due to research that was and continues to be conducted at The Farsight Institute. This is a non-profit research and educational institute based in Atlanta, Georgia, that is dedicated to the continued development of the science of consciousness using remote viewing as the primary research tool. I am the director of the institute. Much of the research that is conducted is available for free on the Internet at the institute's web site, www.farsight.org.

Underpinning all of the research is the hypothesis that all humans are composite beings. This means that we have two fundamental aspects: a soul and a body. In the current jargon of remote viewing, the soul is called the "subspace aspect" of a person. The physical realm of solid matter is both separate from and connected to subspace. Once our physical bodies expire, we are no longer composite beings, and we continue our existence as subspace entities.

While we are composite beings, physical stimuli tend to dominate our awareness. This means that our five senses (taste, touch, sight, hearing, smell) overshadow the more intuitive awareness originating from the subspace side. In practical terms, this means that most people are not aware that they even have a subspace aspect. In short, soul voices are deafened by the din of our five physical senses.

In order to break through this noise, specialized techniques are required. In general, these techniques focus on shifting a person's awareness away from the five physical senses. It is not necessary to force a shift in one's awareness toward the subspace aspect. This happens automatically once a person's awareness is no longer riveted on the physical side of life.

For this reason, I advise combining the practice of remote viewing with the practice of meditation. The form of meditation that I enjoy is Transcendental Meditation (TM), or the more advanced TM-Sidhi Program. My preference is based on the fact that TM is a mechanical procedure, and it has no belief or religious requirement associated with it. The mechanics of TM are also quite stress free and relaxing. Again, these are only my preferences. Many people who participate in other programs for the development of consciousness have also learned remote viewing.

Remote viewing is a natural process of a deeply settled mind. Remote perception works best when it is not forced in any way. I have often said that the ancient seers were our first human astronauts. While in a deeply relaxed state, they let their minds roam across the fabric of the universe, and some perceived what was there with surprising accuracy.

The subspace mind, the intelligence of the soul, perceives and processes information differently from the physical mind. All evidence suggests that the subspace mind is omnipresent across space and time. It is everywhere at once. Using the capabilities of the subspace mind, remote viewing involves no more than shifting one's awareness from one place and time to another. You do not go anywhere when you remote view. You do not leave your physical body. You do not induce an altered state of consciousness. You merely follow a set of procedures that allows you to shift your awareness from one area of your intelligence to another.

As physical beings, though, we must translate the information perceived by our subspace aspects into physical words, pictures, and symbols so that this information can be conveyed to others within the physical realm. Scientific Remote Viewing (SRV) facilitates this translation. Remote viewing would be impossible in the absence of the human soul, since it is otherwise physically impossible for an individual's conscious mind to perceive things without direct physical contact of some sort.

COMMUNICATION WITH THE SOUL

Soul-level communication is not as easy as you might initially think. On one level, communication using the soul is as natural as breathing. While the theoretical principles underlying how this is done are quite simple, knowing with some degree of certainty that the communication is accurate is more difficult.

Subspace information has a mental flavor that is distinctly different from that obtained from the five physical senses. It is much more subtle and delicate. For this reason, sensory input from the five physical senses needs to be kept to a minimum both immediately prior to and during a remote-viewing session. That's why one begins with meditation or other procedures to calm the mind, and then to shift one's awareness away from the physical senses.

The five physical senses are not the only hurdles confronting the remote viewer. The thinking, judgmental, and evaluative processes of the conscious mind can also inhibit success. The conscious mind can contaminate accurately perceived information. The amount of information the conscious mind has regarding the target during the remote-viewing session has to be minimized.

Information coming from the subspace mind is typically called "intuition." This is a feeling about something, which one otherwise would have no direct knowledge of on the physical level of existence. For example, many mothers say they know when one of their children is in trouble. They feel it in their bones, so to speak, even when they have not been told anything specific regarding their child's situation. SRV systematizes the reading of intuition.

Using SRV, the information from the subspace mind is recorded before the conscious mind has a chance to interfere with it using normal intellectual processes such as rationalization or imagination. With nearly all physical phenomena, a time delay exists between sequential and causally connected events.

For example, when one turns on a computer, it takes awhile for the machine to boot up. When the institute teaches remote viewing to novices, we exploit the fact that there is approximately a three-second delay between the instant the subspace mind obtains information and the moment when the conscious mind can react to this information. The subspace mind, on the other hand, apparently has instantaneous awareness of any desired piece of information. In general, the novice viewer using SRV protocols moves steadily through a list of, say, a few hundred things at basically a three-second clip for each one. The tasks carried out in the protocols are carefully designed to produce an accurate picture of much of the target by the end of the session.

It is crucial to emphasize at this point that there must be no deviation from the grammar of the protocols. This is particularly true for novices. If there is a deviation, one only has to be reminded that it is the conscious mind that designs this deviation. When this happens, the subspace mind loses control of the session, and the data from that point on in the session are often worthless.

TARGET COORDINATES

Scientific Remote Viewing always focuses on a target. A target can be almost anything about which one desires information. Typically, targets are places, events, or people. But advanced viewers also work with more challenging targets.

An SRV session begins by executing a set of procedures using target coordinates. These are essentially two randomly generated four-digit numbers that are assigned to the target. The remote viewer does not know what target the numbers represent, yet extensive experience has demonstrated that the subspace mind instantly knows the target even if it is only given its coordinate numbers. The remote viewer is not told the target's identity until after the session is completed.

For all of the remote-viewing sessions presented in this book, the only thing I was given prior to the beginning of the sessions was a fax or an e-mail from my "tasker" telling me the target's coordinates. The tasker is someone who tasks or assigns a target. For example, if the target was the Taj Mahal, I would not be told to remote-view the Taj Mahal, since this would activate all of the information held by my conscious mind regarding this structure, meaning that I would have a difficult time differentiating the remote-viewing data from memories or imagination. Instead, the tasker would tell me that the numbers were, say, 1234/5678. My conscious mind would not know what target is associated with these numbers, but my subspace mind would know the target immediately. A productive session would then include good sketches of the structure, or at least aspects of the structure, together with written descriptive data of the building and its surroundings, including people who may be in or near the building.

THE SRV PROTOCOLS

Scientific Remote Viewing has five distinct phases, which follow one after the other during an SRV session. In each phase the viewer is brought into either a closer or an altered association with the target. SRV is performed by writing, on pieces of plain white paper with a pen, sketches and symbols that represent aspects of the target. The viewer then probes these marks with the pen to sense any intuitive ideas. Since the subspace mind perceives all aspects at once, probing a mark is a way of focusing attention on the desired aspect.

The five phases of the SRV process are as follows:

- *Phase 1.* This establishes initial contact with the target. It also sets up a pattern of data acquisition and exploration that is continued in later phases. This is the only phase that directly uses the target coordinates. Once initial contact is established, the coordinates are no longer needed. Phase 1 essentially involves the drawing and decoding of what is called an "ideogram" in order to determine primitive descriptive characteristics of the target.
- *Phase 2.* This phase increases viewer contact with the site. Information obtained in this phase employs all of the five senses: hearing, touch, sight, taste, and smell. This phase also obtains initial magnitudes that are related to the target's dimensions.
- *Phase 3.* This phase is a sketch of the target.

- *Phase 4.* Target contact in this phase is more detailed. The subspace mind is allowed significant control in solving the remote-viewing problem by permitting it to direct the flow of information to the conscious mind.
- *Phase 5.* In this phase the remote viewer can conduct some guided explorations of the target that would be potentially too leading to be allowed in Phase 4. Phase 5 includes specialized procedures that can dramatically add to the productivity of a session. For example, one Phase 5 procedure is a locational sketch in which the viewer locates a target in relation to some geographically defined area, such as the United States.

CATEGORIES OF REMOTE-VIEWING DATA

Remote-viewing data can be obtained under a variety of conditions, and the nature of these conditions produces different types of data. There are six different types of remote-viewing data, and there are three distinguishing characteristics of the various types of data. The first distinguishing characteristic is the amount of information the viewer has about the target prior to the beginning of the remote-viewing session. The second is whether or not the viewer is working with a person called a "monitor," explained below. The third is determined by how the target is chosen.

Type 1 Data

When a remote viewer conducts a session alone, the conditions of data collection are referred to as "*solo.*" When the session is solo and the remote viewer picks the target (and thus has prior knowledge of the target), the data are called Type 1 data.

Knowing the target in advance is called "*front loading.*" Front loading is rarely necessary and should be avoided in general, but sometimes a viewer simply needs to know something about a known target and has no alternative. Such sessions are very difficult to conduct from a practical point of view. The viewer's conscious mind can more easily contaminate these data, since the viewer may have preconceived notions of the target. Rarely do even advanced viewers attempt such sessions. Any findings are considered suspect, and attempts are made to corroborate the data with other data obtained under blind conditions (see Type 2 data).

Type 2 Data

When the target is selected at random from a predetermined list of targets, the data are called "Type 2" data. For this, a computer (or a human intermediary) normally supplies the viewer with only the coordinates for the target. Even if the viewer knows the list of targets, since sometimes the viewer has been involved in designing the list, only the computer knows which coordinate numbers are associated with each target. It is said that the viewer is conducting the session blind, which means without prior knowledge of the target.

Type 3 Data

Another type of solo, blind session is used to collect Type 3 data. In this case the target is determined by someone (a tasker). During training, viewers may (rarely) receive some limited information regarding the target—perhaps whether the target is a place or an event. Advanced viewers are normally not told anything other than the target coordinates.

Solo sessions can yield valuable information about a target, but trainees often find that more in-depth information can be obtained when someone else is doing the navigation. This other person is called a "monitor," and monitored sessions can be spectacularly interesting events for the new remote viewer.

Type 4 Data

There are three types of monitored SRV sessions. When the monitor knows the target but communicates only the target's coordinates to the viewer, this generates Type 4 data. These types of monitored sessions are often used in training. Type 4 data can also be very useful from a research perspective, since the monitor has the maximum amount of information with which to direct the viewer. In these sessions, the monitor tells the viewer what to do, where to look, and where to go. This allows the viewer to almost totally disengage his or her analytic mental resources while the monitor does all of the analysis.

One of the troubles with Type 4 data for advanced practitioners is that their telepathic capabilities become so sensitive that they can be led during the sessions by the thoughts of the monitors. Even slight grunts, changes in breathing, or any other signal, however slight, can be interpreted as a subtle form of leading by the monitor, which in turn could contaminate the data. To eliminate these problems, advanced monitored sessions are normally conducted under double-blind conditions, yielding Type 5 data.

Type 5 Data

For this level both the viewer and the monitor are blind, and the target either comes from an outside agency or it is pulled by a computer program from a list of targets. Sessions conducted under these conditions by proficient viewers tend to be highly reliable. The disadvantages are that such sessions do not allow the monitor to sort out the most useful information during the session. To address this limitation, scripts are often given to the monitor in advance of the session. These scripts contain no target-identifying information, but they do give clear instructions as to which procedures and movement exercises need to be executed (and in what order).

Type 6 Data

These data come from sessions in which both the monitor and the viewer are front loaded with target information. This type of session was occasionally used when there were very few professionally trained viewers and monitors, information needed to be obtained quickly, and there was no one else available to task with the session. Type 6 data are rarely if ever collected these days.

Descriptions of remote-viewing sessions in this book use Type 2 and Type 3 data. The sessions using verifiable targets in Part II all employ Type 3 data. When I conducted these sessions, I had absolutely no prior knowledge of the targets in any way. For the substantive sessions presented in Section 2, a mixture of Type 2 and Type 3 data are used. I was involved in creating a list of approximately 20 highly varied targets for the Type 2 sessions. I gave the list of targets to an intermediary who mixed them up, assigned random coordinate numbers to each one, and then gave me the coordinate numbers. I viewed all Type 2 targets in a batch before being told the cue/coordinate associations. The Type 3 data used in Section 2 involve targets that were designed by someone other than myself and that were given to me blind.

THE REMOTE-VIEWING EXPERIENCE

When at peace inwardly, and generally stress free, beginners perceive a target with a clarity characteristic of, say, a light on a misty night. While there may be difficulty discerning the precise meaning and distance of a light under such conditions, there is nonetheless no doubt that a light is perceived. With experience and skill, a remote viewer can perceive all sorts of details relating to a target, just as an experienced yachtsman, upon seeing the light, can soon discern the outline of the nearby coast, and the identity of the lighthouse from which the shrouded beacon shines.

Learning how to remote view from a book is not optimal. The primary reason for including these methods is not to teach Scientific Remote Viewing, but to explain it to people who want to understand and interpret remote-viewing data. Students of remote viewing must understand that the effectiveness of any procedures depends not only on the procedures themselves, but also on how well they are executed. This, in turn, depends on the quality of instruction and feedback. In a classroom, regular instructions are directed at a student's work while the initial learning process is under way (and before counterproductive habits are formed). These instructions help obtain the highest level of performance. Nonetheless, many students can achieve a minimal level of effectiveness by systematically studying the procedures presented here without the assistance of classroom instruction.

The term "remote viewing" is actually not entirely appropriate. The experience is not limited to visual pictures. All of the senses—hearing, touch, sight, taste, and smell—are active during the remote-viewing process. More accurate is the term "remote perception." Nonetheless, since "remote viewing" has been widely adopted in the scientific as well as the popular literature, it makes sense simply to continue using the current term.

When one looks at an object, the light reflected off that object enters the eye, and an electrochemical signal is generated that is transmitted along the optic nerve to the brain. Scientific studies have demonstrated that this signal is "displayed" on a layer of cells in the brain, the way an image is projected from a movie projector onto a movie screen. The brain then interprets this image to determine what is being seen. When someone remembers an object, the remembered image of the object is also projected onto that same layer of cells in the brain.*

When remote viewing, one also perceives an image, but it is different from the remembered image or the ocular image. The remote-viewing image is dimmer, foggier, and fuzzier. Indeed, one tends to "feel" the image as much as one visualizes it. The human subspace mind does not transmit bright, high-resolution images to the brain, and this fact is useful in the training process for SRV. If a student states that he or she perceives a clear image of a target, this image almost certainly originates from the viewer's imagination rather than from subspace.

This does not mean that the relatively low-resolution remote-viewing experience is inferior to a visual experience based on eyesight. Remember that all of the five senses—plus the sense of the subspace realm—operate during the remote-viewing process. Thus, it is actually possible to obtain a much higher-quality collection of diverse and penetrating data. The remote-viewing experience is simply different from, not superior or inferior to, physical experience of observation.

*If one remembers an object and visualizes it while the eyes are open and looking at something else, then the same layer of cells in the brain contains two separate projected images. The image originating from the open eyes is the brightest, whereas the remembered image is relatively dim and somewhat translucent, since one can see through the translucent image to perceive the ocular originating image. For those readers who would like to read an accessible but more in-depth treatment of the physiology of visual and remembered images, I strongly recommend an article in the *New York Times* by Sandra Blakeslee titled "Seeing and Imagining: Clues to the Workings of the Mind's Eye," *New York Times*, 31 August 1993, pp. B5N–B6N.

A remote viewer's contact with a target can be so intimate that a new term, "bilocation," is used to describe the experience. Approximately halfway through a session, the viewer often begins to feel he or she is in two places at once. The rate at which data come through at this point is typically very fast, and the viewer has to record as much as possible in a relatively short period of time.

Experience has shown that each viewer is attracted to certain aspects of any particular target, and not all are attracted to the same aspects. One viewer may perceive the psychological condition of people at the target location, whereas another viewer may focus in on their physical health. Yet another viewer may concentrate on the physical attributes of the local environment of the target. For example, I once assigned a target of a bombing to a group of students. One of the students was a doctor and another a photographer. After the session was completed, I reviewed each student's work. The entire class perceived the bombing incident. But the doctor described the physical characteristics of the bombing victims closely, including all of their medical problems resulting from the bombing. On the other hand, the photographer's session read more like a detailed analysis of the physical characteristics of the event, including an accurate description of the geographical terrain where the bombing took place.

Thus, remote viewers go into a session with what they already have—their own personalities. Advanced remote viewers balance these attractions because their training is designed to extract a comprehensive collection of data. But even under the best of circumstances, some level of individual focusing is inevitable for each viewer. For this reason, we use a number of advanced remote viewers for any given project. Each viewer will contribute something unique to the overall results, and a good analyst can put the pieces of the puzzle together to obtain the fullest analysis of the target.

So, you may ask, who should remote view?

In this field there is a distinction between natural and trained remote viewers. Natural remote viewers are generally referred to as "psychics," or when the context is clear, simply "naturals." Naturals typically use no formal means of data acquisition. They simply "feel" the target, and their accuracy depends on how well they can do this. Because naturals may not understand the mechanism by which their talents are achieved, their dependency on the "feel" of the data can cause problems of accuracy. A person's conscious mind can disguise information to make it feel right, when in fact it is not correct at all. Furthermore, since it is difficult to accurately evaluate the "flavor" of psychic data while it is being collected, most naturals have very uneven success histories.

By the end of 1997, The Farsight Institute had trained a large number of people in the basics of Scientific Remote Viewing. With this teaching experience as background, we have identified a clear pattern. Any person of average or better intelligence apparently can be trained to remote view with considerable accuracy. Certain life experiences and educational backgrounds sometimes assist in the process. In week-long introductory classes taught at The Farsight Institute, all or nearly all students have successful remote-viewing experiences, and the instructors generally expect that most sessions conducted after the third day contain some obviously target-related material.

Part of the training process is helping participants identify and interpret subspace-accessed data with increasing precision. All aspects of all targets have a particular "feel." The novice viewers are just beginning to learn what these aspects feel like on an intuitive level.

In addition, Farsight Institute trainees who practice meditation already have a good intuitive sense of subspace. Their initial training moves quickly from learning the mechanics of SRV to the advanced discrimination between complex target characteristics. Meditators often discern new things and have more penetrating and profound remote-viewing experiences more quickly than those who do not meditate. Of course, there are exceptions: many remote-viewing trainees are very good from the start even if they have never meditated.

With this general discussion of Scientific Remote Viewing complete, we are now ready to explain the mechanics of the process and how it works. We begin this in the next chapter by explaining how we identify a target using what is called a "target cue."

Chapter 3

TARGET CUES

Writing an effective target cue is one of the most important criteria in remote viewing. The target cue identifies the target. It is the actual event, person, object, or whatever, that is the focus for a remote-viewing session. Normally, the remote viewer is not told the target cue until after the session is completed. With Type 5 data (double-blind), the monitor also is not told the target cue until after the session is completed.

The initial target cue is given through the target coordinates. Typically, the person who tasks the session has a piece of paper on which the target coordinates and the target cue are both written. In Type 5 data situations, the tasker gives the monitor the target coordinates (normally over phone or fax), and nothing more. Experience has clearly demonstrated that the viewer's subspace mind has instantaneous awareness of the meaning of the target coordinates, and a typical session begins immediately by obtaining information directly related to the target cue.

Humans perceive and process remote-viewing data differently. For example, if someone was told to go into a room and to see what was there, they would need little additional instruction. The request to go into the room and observe is vague, yet most people would not feel uncomfortable with the request, knowing that they would probably be able to sort things out once they got into the room. When they start looking around, they could make an inventory of the room's contents. Their conscious minds would be fully engaged as they entered the room, and most people would perform satisfactorily in this regard even if they had no prior expectations regarding the contents of the room.

With remote viewing, the viewer has minimal help from the conscious mind. The viewer cannot scan everything, evaluate the importance of all that is perceived, make logical choices as to which are the important things to observe, and rank them in order. The remote-viewing experience is more passive; the viewer perceives what is there, but the viewer has only limited evaluative capabilities. Thus, for remote viewing to be most successful, it is necessary to compensate for the relative lack of input from the conscious mind. To do this, one makes the target cue very specific with regard to what is desired from the subspace mind of the viewer.

At The Farsight Institute, we avoid excessively vague cues. For example, if one tasks a target cue of a person (say, just the person's name), then a viewer would be completely accurate if the observed data were anything that related to this person at any time in his or her life. Even a fantasy that the person had during a lunch break would qualify as accurate data. In such a situation, the choice of what to perceive is being determined by the personal preferences of the viewer's subspace mind. To avoid this problem of subjectivity, the instructions in the cue have to eliminate as much ambiguity as possible.

In this chapter I will present one of the more modern forms of cuing that is used at The Farsight Institute. Other cuing forms tend to be more basic versions of that presented here, and readers will see some of these other forms used in later chapters. None are better or worse; they just do different things.

To task a target, one needs a "target definition." A complete target definition has a variety of parts, but they are basically broken down into (1) viewing parameters, (2) the essential cue, and (3) a list of qualifiers.

VIEWING PARAMETERS

Viewing parameters may contain a variety of components. They typically begin with a declaration of the target coordinates. Following this is the essential cue, as it is described below. The target coordinates and the essential cue are placed at the top of the cue so that analysts who sort through large stacks of targets can identify a target by glancing at the top of the page.

Following the essential cue are two primary viewing parameters. The first is the target range. This gives general instructions as to the type of information that is permissible in the session. For example, the range typically limits the target data to only tangibles and intangibles that exist in the target. At first this may seem obvious. However, all targets bleed into other areas, and it is easy for the subspace mind to follow these smears in the data boundaries. For example, the target may be a specific person on a beach on the equator at a given point in time. But that person may be thinking about an Eskimo hunting a polar bear in the Arctic. If a viewer pursues this perception, the viewer may describe polar bears on the beach.

Then comes the second viewing parameter. This specifies the time frame of the target. Many experiments have verified that there is a complete continuum of existence with an infinite number of time lines, both past, present, and future. The subspace mind is equally capable of perceiving all of these. Thus, it is necessary to request the subspace mind to locate targets as they may exist in time frames and realities that are closely connected to our present. Following the second viewing parameter is the target cue, which includes the essential cue and the qualifiers.

THE ESSENTIAL CUE

The essential cue is normally a simple statement or sentence that describes the basic core of the target. The essential cue is both simple and direct. Sometimes a segmented structure is used in writing the essential cue. The cue has multiple parts, with each being separated by a slash (/). The first part of the essential cue is called the "primary cue." The primary cue is the major identifier of the target. Everything that follows is a refinement of this primary identifier. Thus, if the target is a known place or person, the first part must be the name of the place or person. The primary cue is then followed by a slash and one or more secondary cues (each separated by a slash) if greater refinement of the target is required. The cue "event" is sometimes used as the *final* secondary cue to focus a remote viewer on activity at the target. Specific temporal identifiers follow the primary and secondary cues and are placed in parentheses. As a general rule, each target must have one primary cue, and nearly all targets have at least one secondary cue (as needed) as well as a temporal identifier. The format of the essential target cue is as follows:

> *Primary Cue / First Secondary Cue / Second Secondary Cue (Temporal Identifier)*

The following are some examples of essential target cues that follow the segmented format.

Example 1
Napoleon Bonaparte / Battle of Waterloo / event (1815)

Example 2
John F. Kennedy assassination / event (22 November 1963)

Example 3
Nagasaki / nuclear destruction / event (9 August 1945)

Effective essential cues must begin with a known, not a conclusion. Errors in cue construction usually result from placing an analytical conclusion in the cue itself. The purpose of a remote viewing session is to gather data for known events so that conclusions can be made during the subsequent analysis of the data. For example, a poorly written essential cue that contains a conclusion would be: "John F. Kennedy assassination / conspiracy." In this cue, one is assuming that there is a conspiracy in the assassination. With remote viewing, one must construct a case for a conclusion based on observable data. If there was a conspiracy in the J.F.K. assassination, this must be established from the data of events and people, not by cuing on the idea of conspiracy.

Since remote viewing always obtains descriptive information about people, things, and events, the conscious mind must later make conclusions based on information supplied by remote-viewing data. For example, a remote viewer could be tasked the J.F.K. assassination (that is, the event itself). The viewer could then be given various movement exercises and cues to obtain as complete a collection of data as possible. In the analysis that follows the remote-viewing session, the analyst can then examine the data for any evidence of a conspiracy. For instance, the data may show more than one source of bullets in the event. But one cannot go into a session assuming that there will be more than one source of bullets. That would bias the data-collection process. Restating this important principle, data are collected using neutral target cues, and all analytical conclusions must be made after the data-collection process is completed.

Another example of a poorly written essential cue is: "How to live happily with friendly extraterrestrial neighbors." Many people think that remote viewing can be used to resolve such targets directly. Yet it must begin with a known person, place, thing, or event. A cue about extraterrestrial neighbors would assume the existence of extraterrestrials. At best, one would have to begin with a known, such as an actual sighting of an unidentified flying object, perhaps one documented with a photograph. The remote viewer would then be able to target the object, try to move inside the object, and observe extraterrestrials flying the craft. The viewer would also be able to move into the minds of the extraterrestrials to find out if they are friendly toward humans. With this information, an analyst would have at least something to work with regarding the possibility of friendly coexistence for humans and extraterrestrials.

In general, remote viewing is descriptive. It does not label things, analyze situations, make conclusions, nor does it employ logic or reasoning during the session. For example, if the target is a checkers game, the remote viewer would describe the board, perhaps even drawing the checkerboard pattern in a sketch. The viewer may even correctly place some pieces on the board, and identify the colors of the pieces. But the viewer may not realize during the session that the target is a checkers game. After the session is completed, the analyst can examine the data and conclude that the data seem to correspond with a checkers game.

The target cue has to focus on these descriptive capabilities of remote viewing.

The Qualifiers

Following the essential cue is a list of qualifiers, usually marked with bullets. The qualifiers are written in phrase or sentence format, and they are clear descriptions of specific things that the viewer is supposed to observe and describe. The qualifiers must address the primary goals of the cue, including instructions to observe activity that may be taking place at the target location.

For example, if the cue is a military battle, the qualifiers should explicitly state that the viewer is to observe the battle itself. Otherwise a viewer may perceive what amounts to an inventory list of things and people that are at the scene of the battle, but miss the actual fighting, the sounds of the passing cannonballs, the thunder of the bombs, the shouts of the soldiers, etc. Readers are encouraged to closely examine the qualifiers for the example target cues listed below to obtain a solid sense of what's required. Versions of some of these targets have been used in the actual training of many advanced viewers at The Farsight Institute.

One Complete Example

Target Definition for Target 3292/9537

Essential Cue (and Viewing Protocols): Mike Tyson–Evander Holyfield Championship Boxing Match (28 June 1997). (ESRV)

Viewing Parameter 1: Target Range

The viewer perceives only the intended target as it is specified by this complete target definition. The viewer describes only tangibles and intangibles that exist in this target.

Viewing Parameter 2: Target Links

If the target resides outside of a past, present, or future connection to the temporal and/or spatial reality of the current tasking time frame, then the viewer remote views the target as it exists in its own reality.

If the target time is the moment of tasking, then the viewer remote views the target as it exists in the same temporal and spatial reality of the tasker at the moment of tasking.

If the target time is prior to the moment of tasking, then the viewer remote views the target as it exists in the temporal and spatial reality of the time stream that directly evolves into the temporal and spatial reality of the tasker at the moment of tasking.

If the target time is in the future of the moment of tasking, then the viewer remote views the target as it exists in the most highly probable temporal and spatial reality as it may evolve from the temporal and spatial reality of the tasker at the moment of tasking, given both the existing conditions of the tasker's reality at the moment of tasking, as well as directions for extrapolation into the future if such are specified in the target cue.

Target 3292/9537

Protocols used for this target: Enhanced SRV

The Mike Tyson–Evander Holyfield Championship Boxing Match (28 June 1997). In addition to the relevant aspects of the general target as defined by the essential cue, the viewer perceives and describes the following target aspects:

- Mike Tyson and Evander Holyfield
- the target activity in the boxing ring
- the activity surrounding the boxing ring
- the building within which the target is located
- the thoughts of the people watching the fight inside the building where the match occurs

Examples of Essential Cues with Qualifiers

Target 9148/5716

Madeleine Murray O'Hare / current location. In addition to the relevant aspects of the general target as defined by the essential cue, the viewer perceives and describes the following target aspects:

- the current physical characteristics of Madeleine Murray O'Hare
- the current physical condition of Madeleine Murray O'Hare
- the surrounding environment and current location of Madeleine Murray O'Hare's physical body

TARGET 3985/3159

The Apollo 11 landing on the Moon / event (20 July 1969). In addition to the relevant aspects of the general target as defined by the essential cue, the viewer perceives and describes the following target aspects:

- the actual landing event in which the lander contacts the lunar surface
- the activity of Neil Armstrong as he emerges from the lunar lander and walks on the lunar surface for the first time
- Neil Armstrong planting the U.S. flag on the lunar surface

TARGET 6459/3395

Ted Bundy's execution / event. In addition to the relevant aspects of the general target as defined by the essential cue, the viewer perceives and describes the following target aspects:

- Ted Bundy during the execution event
- Ted Bundy's surroundings during the moment of execution
- the people near Ted Bundy during the execution
- the emotions of Ted Bundy as well as the emotions of the people near him who are watching the execution
- the method by which the execution is performed

Here is an esoteric target. Before giving an esoteric target with an extensive list of qualifiers, the tasker must have some information strongly suggesting that such a target in fact exists. Such information can come from more open-ended cues.

TARGET 3292/9537

The living physical subjects and their facilities that are currently located on Mars (at the time of tasking). In addition to the relevant aspects of the general target as defined by the essential cue, the viewer perceives and describes the following target aspects:

- the physical environment of the subjects' living conditions
- the age and gender variations among the subjects
- the emotional state of the subjects

- the dominant groups among the subjects, including any governmental organizations
- the primary thoughts of the collective consciousness of the subjects
- the level of technology available to the subjects

Chapter 4

Phase 1

THE PRELIMINARIES

1. Consciousness-Settling Procedure

The single most important step needed to obtain a profound remote-viewing experience is a deeply settled mind. For this reason I recommend that remote viewers meditate regularly. While I personally practice Transcendental Meditation (TM), other forms of meditation may be useful as well. Additionally, since a settled mind is so essential to deep target penetration, the practice of SRV begins with a procedure that helps to settle the mind in an appropriate fashion. This practice is called the SRV "Consciousness-Settling Procedure" (or CSP), and it is composed of a few simple techniques commonly practiced in a number of meditation traditions.

CSP must be done immediately prior to each SRV session by both the viewer and the monitor. CSP takes approximately 15 minutes total. In Type 4 and Type 5 settings, monitors and viewers need to communicate 15 minutes before each session to coordinate the precise timing of the beginning of the SRV session. Here are the steps for CSP:

1. Sit comfortably in silence with the eyes closed for 30 seconds.
2. Perform a brief body massage. (Some meditation traditions recommend that the massage be executed slightly differently for men and women, and I describe these recommendations here. I am not clear as to why these gender-related differences exist, or if the need for the differences is real.) The massage begins by gently pressing the hands against the face, then upward on the top of the head, back down the neck, and toward the heart. (All massage elements move toward and finish at the heart.) Then men continue by gently using the left hand to press and massage first the right hand, and then up the arm, and back down toward the heart. Again, this is all done with the left hand. Women do the same, but they begin by massaging the left hand and arm (back toward the heart) with the right hand. Then both men and women switch arms and massage the other hand and arm, again, back toward the heart. Then men continue by massaging the right foot and leg, upward toward the heart. This is done with both hands pressing gently. Then massage the left foot and leg, again, upward toward the heart. Women do the same, but they begin with the left foot and leg, upward toward the heart, before repeating the process for the right foot and leg. This is best done with the eyes closed. Total time for the massage is about a minute.
3. While sitting comfortably with the back straight, perform a breathing technique that is called "pranayama." Begin with 10 seconds of fast pranayama. This is done using very short, gentle breaths, closing one nostril at a time after each outward and inward breath. Close the nostrils (one at a time) with the thumb and the middle fingers (alternately) of one hand. Men use their right hand to do this while women use their left. The mechanics of the procedure are similar to slow pranayama (see below), except that the breaths are very short and rapid (although still gentle). This is best done with the eyes closed. The procedure should be effortless and easy, and if someone is experiencing any problems like dizziness or hyperventilation, it is being performed incorrectly and its practice should be

discontinued until getting personal instruction in this technique.

4. While sitting comfortably with the back straight, perform 9 to 10 minutes of slow pranayama. This is done similarly as with the fast pranayama, but using normal breaths (not short or long ones), closing one nostril at a time after each outward and inward breath. Be sure to complete both the outward and inward breath before switching nostrils. On the exhaling breath, let the breath flow out naturally, not forcing it. The inhaling breath should take about half the time as the exhaling breath. Hold the breath after inhaling for a brief moment (a second or two) while alternately closing the other nostril with the other finger, and prepare to exhale. The entire procedure should be effortless and gentle. If you feel you need more air, simply take deeper breaths, but do not hyperventilate. You should be breathing normally, just alternating nostrils after exhaling and inhaling. This is best done with the eyes closed.
5. Sit quietly and comfortably for 5 minutes with the eyes closed.
6. Open your eyes and immediately begin the SRV session.

2. Physical Considerations to Beginning the SRV Session

A remote-viewing session begins with a viewer sitting at a clean desk. Ideally, the only items that should be on the desk are a pen and a thin stack of white paper. We use a ballpoint pen with liquid black ink. A good quality pen that does not produce much friction when writing is best. Traditional ballpoint pens that use gummy ink require too much downward pressure when writing.

The ideal training room is neutral in color. Light gray, powder blue, or light brown are suitable colors. It is probably not a good idea to use, say, a child's playroom that has lots of primary colors on the walls. The idea is to minimize the strong stimuli that come in through the senses, such as bright visual colors.

Before remote viewing, a person should be well rested. This cannot be emphasized enough. Tiredness dulls the conscious mind, and a tired conscious mind has difficulty perceiving information originating from the subspace mind. A good night's sleep is ideal for a morning remote-viewing session, and a midday 15–30-minute rest often refreshes one sufficiently for an afternoon session.

One should be comfortably fed before remote viewing. This means that one should not be hungry, and one should also not be overfed. Hunger and feeling stuffed produce physical stimuli that are difficult for the conscious mind to ignore. Remember that the subspace mind yields a relatively weak informational signal to the conscious mind. Try to minimize any physiological stimuli that could swamp the subspace signal.

Remote view in a quiet environment. If possible, close the windows and doors of the remote-viewing room. Also turn off the ringer of the phone for the time that it takes to complete the session. Turn off any radios or televisions that may be audible nearby.

Avoid wearing any perfume, cologne, aftershave, or other strong scents. This is particularly important when training in a group environment. If a viewer is a smoker, it would be best if this viewer wore freshly washed clothes during the session that do not smell of smoke.

People who use recreational drugs, or any other drugs with psychoactive qualities, should not remote view at all. These drugs tend to release any controls that the conscious mind has over the imagination, which is exactly opposite that which is required for successful remote viewing. With respect to drugs of any type, one should try to be as drug free as possible. Individuals who use doctor-prescribed antidepressants should probably not spend much effort trying to remote view. Such antidepressants suppress the nervous system to such a degree that accuracy in remote viewing is highly compromised. Yet individuals using any drugs prescribed by their doctors should not discontinue their use unless directed to do so by their doctor. Learning how to remote view is not as important as maintaining one's health and mental balance.

Before beginning the session, you should sit comfortably on a chair at your desk with both feet on the floor. The legs should not be crossed. You should sit up straight, not off to one side, or sitting on one foot in a lotus position. The hands should be relaxed, with the pen held over a single clean sheet of paper. The paper is positioned in portrait mode (vertically). The stack of paper should be on the viewer's right side of the desk.

THE SRV AFFIRMATION

The SRV Affirmation is normally read aloud with a soft voice, even in solo sessions. The affirmation produces a subtle shift in the sensitivities of the mind that helps to connect the awareness of the conscious mind to the perceptive capabilities of the subspace mind. The SRV Affirmation is designed to closely approximate the way sequential, connected thoughts are felt telepathically, piece by piece, one "thought-ball" at a time. Viewers should read the affirmation slowly, pausing briefly after each comma or period. Here is the SRV affirmation:

SRV Affirmation

I am a spiritual being. Because I am a spiritual being, I am able to perceive beyond all boundaries of time and space. My consciousness is ever present with all that is, with all that ever was, and with all that ever will be. It is in my nature, as a human, to be able to perceive, and thus to know, all that there is to know. Everywhere, at all times, I seek to learn, and thus to evolve. To further my own personal growth, and to assist others in their growth, I direct my attention to a chosen point of existence. I observe what is there. I study it carefully. I record what I find.

THE HEADER

Next, write the SRV identifying header on the top of the first piece of paper. The viewers declare the condition of their physical state (PS), their emotional state (ES), or any advanced perceptuals (AP) centered at the top of the first page. Declaring PS and ES let the conscious mind account for your physical and emotional states, thereby releasing any psychological pressure that could be present. These declarations can be positive, neutral, or negative. Positive declarations include, "I really have a happy glow this morning," or anything else that is upbeat. Negative declarations include having a sore foot, or being upset with the quality of lunch. Unusually strong PS or ES declarations, such as just having had a fight with a spouse, may suggest that the session might be postponed until later. Similarly, if one is in significant pain due to, say, severe arthritis, it might be better to delay the session until the pain abates.

In some ways it is useful to compare the conscious mind to the mentality of a small child. When the conscious mind is experiencing something, it likes to be heard. Declaring the PS and the ES satisfies this need. This helps the conscious mind relax, circumventing its natural desire to force the issue of having its needs recognized later in the session, potentially corrupting the integrity of the data.

Often a viewer begins a session thinking that he or she has an idea as to what the target is. Such ideas are advanced perceptuals, and any thoughts along these lines need to be declared at the outset, or they will build in pressure in the conscious mind during the session, and are likely to emerge in some form during the actual data flow. Declaring these APs in advance again relaxes the conscious mind by satisfying its desire to be heard, thereby minimizing the risk of contaminating the data.

To the right of the PS, ES, and AP is the identifier of the remote viewer. At The Farsight Institute we use a code called a viewer identification number (VIN), but a name would do just as well. Below the name or viewer identifier is the date written in the U.S. military or European format (day/month/year). Below this is the beginning time of the remote-viewing session.

To the left of the page is the data type, and below that is written the monitor's name or identification number (MIN—if the session has a monitor). To summarize, the format of the initial header is as follows:

Type 4	PS—I feel fine.	VIN
MIN	ES—OK, very settled	7 September 1995
	AP—None	11:33 a.m.

Readers are encouraged not to perceive this initial header as a frivolous formality. Everything is carefully structured in SRV. Following these details from the outset of the session focuses the attention of the conscious mind on the structure of the page. Further, trainee viewers should follow all of the seemingly petty structural details of these protocols, including formatting issues involving indentations, dashes, and colons. Once a remote-viewing session is proceeding at a fast speed, the conscious mind can do little else but keep track of these structural details. This frees the informational conduit of the subspace mind from the controlling influence of the conscious mind. Figuratively, this ties the hands of the conscious mind with activity, allowing the subspace mind to slip the data past the conscious mind with minimal interference.

THE IDEOGRAM

After saying the SRV affirmation, the viewer receives the target coordinates from the monitor. The monitor makes sure to speak deliberately and clearly so that all the numbers can be heard. The target coordinates are two four-digit random numbers, and the monitor places a slight pause between the two groups of numbers. On the left side of the page, the viewer writes the first four-digit number, then the second four-digit number directly under the first.

After writing the target coordinates, the viewer immediately places the point of the pen on the paper to the right of the coordinates. At this point an ideogram is drawn. An ideogram is a spontaneous drawing that takes only a moment to complete. The pen does not leave the surface of the paper until the ideogram is completed. Ideograms normally are simple, but complex ideograms can occur. In general, each ideogram should represent one (and only one) aspect or "gestalt" related to the target. For example, if the target is near a body of water, an ideogram could represent water. If there is an artificial structure at the target site, another ideogram could represent this structure, and so on.

Only one ideogram is written for each recitation of the target coordinates. In Phase 1, the monitor usually recites the target coordinate numbers three to five times, enabling the viewer to draw and decode a few ideograms, thereby obtaining information relating to different target gestalts. Each time the viewer writes down the target coordinates, it is said that he or she is "taking" or "receiving" these coordinates.

After drawing the first ideogram, the viewer then writes the capital letter "A" followed by a colon to the right of the ideogram. The viewer then describes the movement of the pen while writing the ideogram, writing this all down after the "A:". The description must describe the process of the pen's movement without the use of labels. The following words are generally acceptable in this regard: vertical upward, vertical downward, diagonal upward, diagonal downward, sloping (upward or downward), curving (upward or downward), moving (upward, downward, or across), slanting (upward or downward), curving over, curving under, horizontal flat across, horizontal flat along, angle. Words ending in "ing" or "ward" are generally preferred. Labels such as "a circle," "a loop," or "a square" are to be avoided. Labeling adds conceptual meaning to data in remote viewing, and that is conscious-mind analysis. All of remote viewing is built upon perceptions that begin at the lowest level of conceptual abstraction and gradually move to higher levels of abstraction. In the beginning of Phase 1, the lowest level of conceptual analysis is required.

PROBING THE IDEOGRAM

This is a delicate matter. The viewer places the point of the pen on the ideogram itself and gently (but firmly) pushes the pen downward (into the table). The novice viewer can probe one or more times but should avoid more than four attempts. Each probe lasts between one and two seconds (no longer than three seconds). While the pen is in contact with the line, the viewer normally perceives some feeling about the target. Too brief a contact does not allow the nervous system to register the impression sufficiently to allow for accurate decoding. Too long a contact allows the conscious mind to intervene in the process and distort or fabricate the data. After the probe, the pen is removed from the ideogram, and the viewer searches for a word to describe the sensation that was perceived during the probe.

The first time that the viewer probes the ideogram, the attempt is made to discern what is called a "primitive descriptor," of which there are six possible choices, with one exception. These are: hard, soft, semi-hard, semi-soft, wet, or mushy. While probing the ideogram, the viewer will actually sense the pen moving into the paper and table if the target is soft, wet, or mushy. Although this seems logically impossible due to the firmness of the writing surface, it nonetheless is consistently perceived by viewers. When gently pushing the pen into the paper, it will also feel wet if the target has water. The viewer must choose only one of the six possible descriptive options given above. No substitutions should be made, since this would invite the conscious mind to enter the process more fully. The choice of primitive descriptors is then written under the written description of the movement of the pen.

The one exception to picking one of the six primitive descriptors is if the viewer perceives movement or energetics in the ideogram. If this occurs, the viewer may or may not also perceive one of the six primitive descriptors. If the viewer does, then the chosen descriptor is declared and the viewer proceeds with the next step. However, if you perceive only movement or energetics, abandon the attempt to perceive a primitive descriptor and move directly to declaring an advanced descriptor.

After obtaining a primitive descriptor, the viewer probes the ideogram again to obtain what is called an "advanced descriptor." There are five choices, and the viewer must use only one of these choices. These are: natural, man-made, artificial, movement, energetics. After probing the ideogram, the viewer writes the advanced descriptor under the primitive descriptor.

Readers should note that there is a difference between "man-made" and "artificial." While everything that is man-made is artificial, not everything artificial is man-made. For example, a beaver dam is artificial, but it is not man-made. Note also that energetics refers to a feeling that the target is associated with some significant quantity of energy. This energy can be in any form: kinetic, radiant, explosive, etc. While movement can also indicate an expenditure of energy, the movement of a snail or a slowly driven car might not be perceived as energetics.

Underneath part A, the viewer writes "B" followed by a colon. The viewer then declares what he or she perceives the ideogram to represent. The most common declaration is "No-B." While you must have one primitive descriptor and one advanced descriptor per ideogram, you do not have to declare a substantive B. However, the viewer must at least write "No-B."

For B, there is no fixed list of possible declarations. To assist students, however, we offer a list during the first few days. The list is: No-B, structure, water, dry land, wet land, motion, subject, mountain, city, sand, ice, swamp.

Note that these declarations are at a higher level of abstraction than when describing the movement of the pen while drawing the ideogram. The entire process in Phase 1 moves from lower to higher levels of abstraction as follows: describing the movement of the pen, primitive descriptors, advanced descriptors, and an interpretive declaration of the meaning of the gestalt. Yet the viewer must remember that the declaration that is made in part B is still very low-level. For example, a viewer could not declare that the gestalt represents an automobile, a computer, a skyscraper, or a spaceship, since these declarations would be far too high-level, involving conscious-mind interpretations that greatly exceed the quality and quantity of data that are available at this point in the session. For example, if the target really is a skyscraper, then the best that could be determined at this point is that the target is associated with a structure.

Following the declaration of B, the viewer writes "C:" followed by the viewer's intuitive perceptions about what the ideogram feels like. This is usually just a word or two that describes very low-level perceptions relating to the ideogram. Examples of such perceptions are colors or textures (such as rough, smooth, polished, etc.). The viewer may also feel the perception of size, such as big or small, short or tall, wide or narrow. A viewer may also write "No-C" if the previously declared data capture all of the ideogram's nuances.

To summarize, the Phase 1 procedures are (1) take or receive the target coordinates, (2) draw an ideogram, (3) describe the movement of the pen during the drawing of the ideogram using process terms rather than labels, (4) probe the ideogram for primitive descriptors, (5) probe the ideogram for advanced descriptors, (6) make an initial declaration of a low-level description of the target aspect that is captured by the ideogram, or simply state that there is no declaration (i.e., No-B), and (7) list other intuitive feelings regarding the ideogram, if there are any.

This entire sequence is typically done three to five times in Phase 1 (going through all seven steps each time). The idea is not to use Phase 1 to identify all of the aspects of the target, but rather to establish initial contact by describing a few of the primary target aspects only. The viewer then proceeds immediately to Phase 2.

One final note about the ideograms: if an ideogram is not decoded correctly, it is nearly always immediately repeated with the next taking of the coordinates. Thus, a self-correction factor is built into the Phase 1 procedures. If an ideogram returns subsequent to a different ideogram emerging from a different taking of the coordinates, this indicates that the initial ideogram was decoded correctly previously, and that most or all of the primary gestalts have been properly expressed. After decoding a repeating ideogram, the viewer moves on to Phase 2.

For example, let us say that the first ideogram is decoded as a structure. The second ideogram looks different, and from this we assume that the first ideogram was decoded correctly. We decode the second ideogram saying that it is hard and natural, with a B: of "land." On the third taking of the target coordinates, the second ideogram returns. This tells us that we most likely made a mistake in decoding something in the previous (second) ideogram. We probe again, this time finding that the ideogram really feels more like it is hard and man-made. We declare "No-B." We take the coordinates again and the structure ideogram returns. Now we know that we have exhausted all of the major gestalts. We then decode the final ideogram and move on to Phase 2. After the end of the session, we find out that the target was a shopping mall containing a structure and a large parking lot (that is, man-made land).

IDEOGRAM DRILLS

Students need to develop skill in drawing ideograms. Practice and some drills are required. Our students typically drill with a few standard ideograms that have established meanings. They are "established" because many viewers use these same ideograms to represent the same things. Usually seven or eight pages of drills are all that is required to set in place the initial ideogram vocabulary. In the drill, an instructor repeats words like "structure," and the student quickly draws a structure ideogram. Common ideograms that are useful for drill purposes are presented in Figure 4.1.

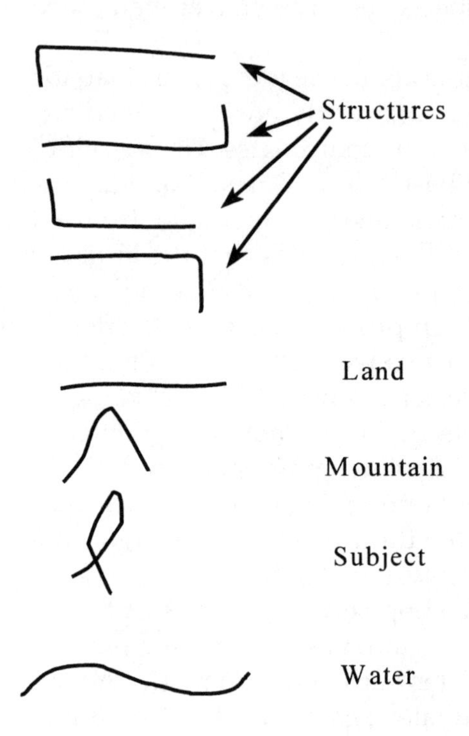

Other ideograms are developed individually for each student. Such ideograms do not have a set pattern, and may vary widely from person to person. Ideograms for such things are drilled not by telling the student what the ideogram looks like, but by just repeating the gestalt (such as the word "movement"), allowing the student to draw whatever comes naturally. The ideograms typically settle down into a set pattern for each gestalt after only a few repetitions. "Person" or "subject" ideograms are often very individualistic in this regard. As a result of these drills, most students develop a minimum of five or six distinct patterns in their ideogram vocabulary. Should a student ever develop an "ideogram rut," in which all ideograms always look alike, then 10 minutes of drill using a variety of ideograms usually fixes this problem.

DEDUCTIONS

What do you do if the conscious mind makes a high-level guess as to the identity of the target or target fragment? This is called a "deduction." A deduction has two components. First, it is a conclusion (as in "to deduce") that the conscious mind makes regarding the target. The conscious mind is basically watching the data flow between the subspace mind and the physical body (the hand holding the pen). The conscious mind needs very little information before it leaps into the process with a guess as to the meaning of the data. This conclusion may indeed be correct, but the viewer cannot know until the target identity is revealed at the end of the session. Thus it is important to remove the conclusion from the data recording process, which leads to the other half of the meaning for "deduction." A deduction is also a subtraction from the data flow. If this high-level conclusion is removed from the data collection, it will not contaminate the remainder of the data flow.

Nearly all deductions describe some true aspect of the target, but a remote viewer doesn't know during a session what that aspect is. For example, if a target is the destruction of the Hindenberg blimp, it follows that kite, balloon, fireworks, and TWA Flight 800 could all be deductions. The idea of a kite captures the notion that the Hindenberg flew, the balloon reflects the structure of the blimp, fireworks reflect the explosion that resulted in the destruction of the Hindenberg, and TWA Flight 800 identifies the idea that an airborne vehicle carrying passengers exploded causing loss of life.

Do not worry about the inaccuracies inherent in deductions. Remember, deductions are not remote-viewing data. They are guesses made by the conscious mind, nothing more. However, deductions can be very useful when analyzing the data afterward. Deductions can convey meaning about a target that is difficult to express. For example, someone could be remote viewing a slave labor camp during the time of the Pharaohs, and give Auschwitz as a deduction. Such a deduction has many parallels with the actual target. Jews were the subjects of slavery, repression, misery, and death in both settings. But more important, the analyst may be alerted to the magnitude of the misery that was experienced in Egyptian slave labor camps through the deduction of Auschwitz. This could be useful in interpreting the remainder of the session should the viewer describe extreme levels of suffering among the actual target subjects.

Regardless of the potential accuracy of deductions, they must be eliminated from the flow of the data. To accomplish this, the viewer writes a capital letter "D" followed by a dash and the description of the deduction on the right-hand side of the paper. Thus, the deduction mentioned above would be written as "D-Auschwitz." Following this, the viewer must put the pen down on the table for one or more seconds. This action of putting the pen down breaks the flow of the data from the subspace mind, thereby allowing the impression that was made on the conscious mind to dissipate. After a few moments the viewer picks up the pen and continues with the session.

Chapter 5

Phase 2

Phase 1 initiates contact with the target. Phase 2 deepens that contact by systematically activating all of the five senses: hearing, touch, sight, taste, and smell. In Phase 2, viewers write down various cues as well as their initial impressions of these cues. In early training (the first three days), these steps are performed slowly so that students can commit the mechanics of the process to memory. Once this is done, the speed of these steps dramatically increases.

Phase 2 begins by writing "P2" centered at the top of a new sheet of paper. In general, all phases must begin with a new sheet of paper regardless of how much space is left on the previous piece of paper. The page number is entered on the upper right corner of the new page.

The viewer begins by writing the word "sounds" followed by a colon on the left side of the page. Immediately after writing this, the viewer normally perceives some sense of sound, although this is obviously not a physical perception. To assist the new viewer in building a vocabulary for this phase, the instructor often recites a list of sounds from which the viewer can choose one or more. This list includes the following: tapping, musical instruments, laughing, hitting, flute, whispering, rustling, whistling, horn, clanging, voices, drums, barking, humming, beating, trumpets, vibrating, crying, whooshing, rushing, whirring. The viewer will often perceive a variety of sounds, and should record all of these perceptions as rapidly as possible.

The viewer then cues on textures that are associated with the target. This is done by writing the word "textures" on the left side of the page, followed by a colon. While writing the cue or immediately afterward, the viewer will sense certain textures and write them down after the colon. To help students during the first few days of training, the following list of textures is read: rough, smooth, shiny, polished, matted, prickly, sharp, foamy, grainy, slippery, wet.

The next sensation is temperature. The viewer writes the abbreviation "temps" on the left side of the page, followed by a colon. As before, one or more temperatures will be perceived immediately, and the viewer must write these down following the colon. The list of possible temperatures that is read to the beginning student is: hot, cold, warm, cool, frigid, sizzling.

The viewer then cues on visuals. These have three components. To begin, the viewer writes "visuals" on the left side of the page followed by a colon. Dropping down and indenting, the viewer writes "colors" followed by a dash (not a colon). The list of colors that is read to the viewer is: blue, yellow, red, white, orange, green, purple, pink, black, turquoise (and others). The viewer may write down colors from this list, or may perceive other colors. In any case, the list is no longer read after the first few days.

On the next line, also indenting as with colors, the viewer writes "lum" for luminescence. As with colors, the cue is followed by a dash, not a colon. The list of possibilities is: bright, dull, dark, glowing.

The final visual is contrasts. This cue is written under "lum," and is followed by a dash. The list of possible contrasts is: high, medium, low.

Dropping down again, but now returning to the left side of the page (that is, no longer indented), the viewer cues on tastes. This is done by writing the word "tastes" followed by a colon. The list of possible tastes is: sour, sweet, bitter, pungent, salty.

The final cue for the five senses is smell. The viewer writes the cue "smells" on the left side of the page followed by a colon. As with all other cues, the viewer will immediately perceive some smells, and these must be recorded without delay. The list of possible smells is: sweet, nectar, perfume, flowers, aromatic, shit, burning, dust, soot, fishy, smoke (also cold and hot).

After recording the data from the five senses, the viewer is normally drawn much closer to the target. Evidence of this is that the viewer almost always perceives many magnitudes of the target. Most magnitudes are essentially quantities. They tend to answer the question of "How much?"

To probe for these target aspects in Phase 2, the viewer first indents on the page and writes "Mags" followed by a colon. Dropping down and indenting further, the viewer cues on the various types of magnitudes shown in the following list. The viewer should not write down the cues for the magnitudes, since these cues are long and this could dangerously slow down the recording of the data.

Here is the list of cues and a collection of possible choices. Advanced viewers typically develop a larger vocabulary of descriptive magnitudes.

[VERTICALS] high, tall, towering, deep, short, squat
[HORIZONTALS] flat, wide, long, open, thin
[DIAGONALS] oblique, diagonal, slanting, sloping
[TOPOLOGY] curved, rounded, squarish, angular, flat, pointed
[MASS, DENSITY, SPACE, VOLUME] heavy, light, hollow, solid, large, small, void, airy, huge, bulky
[ENERGETICS] humming, vibrating, pulsing, magnetic, electric, energy, penetrating, vortex, spinning, churning, fast, explosive, slow, zippy, pounding, quick, rotating

The viewer must perceive magnitude data for at least three of the six dimensions before proceeding further. If the viewer fails to perceive data for at least three, the viewer is undoubtedly editing out data.

In the beginning of training, a viewer sometimes claims not to perceive anything. This is almost always a matter of editing out data, which occurs when the conscious mind enters the remote-viewing process and makes a decision that a piece of data cannot be correct. This is usually perceived as doubt in the mind of the remote viewer.

To remedy this, an instructor encourages the student not to edit out anything, and to write down the data immediately. This raises an important point. It does not matter how the conscious mind is occupied as long as the viewer stays within the structure of the remote-viewing protocols. This means that the viewer need only keep track of what is to be done next, and to mechanically perform that duty correctly.

DECLARING THE VIEWER FEELING

At the end of recording dimensional magnitudes, the viewer begins to perceive aspects of the target very strongly. These aspects could be anything: emotional, physical, or whatever. When this happens, the viewer's conscious mind responds to the data, and this response must be declared in order to limit its ability to contaminate the data not yet collected. This response is called a "viewer feeling," and it is declared by writing the letters "VF" followed by a dash, and then the declaration of the feelings of the viewer. The viewer's feeling is *not* the viewer's perception of the target. Rather, it is the viewer's gut response to the target.

The viewer *must* have a viewer feeling at the completion of the initial pass through Phase 2, but it is not required or even desired that the viewer feeling be dramatic. The viewer's gut response can be simply, "OK," if that is how the viewer feels at that point. A list of common examples of viewer feelings is: I feel good, disgusting, I feel happy, interesting, awful, this place stinks, this is gross, I feel light and lifting, I feel spiritual, enlightening, wow! The most important thing to remember about the viewer feeling is that it is not data. It does not describe the target. It describes the viewer's emotional response to the target. By declaring the viewer feeling, we acknowledge it and remove it from the data flow.

After declaring any viewer feeling, the viewer must put the pen down momentarily, letting the feeling dissipate before picking up the pen again and continuing with the session. In this regard, a viewer feeling is treated similarly to a deduction.

Chapter 6

Phase 3

Phase 3 consists of drawing a sketch guided by the intuitive feelings of the viewer. These can be spontaneous sketches of the target, but they also can be somewhat analytical, based on what was perceived earlier in the session. The sketches can sometimes be detailed, graphical representations of the target, but often they are more like pictorial symbols, partially descriptive but also symbolic of the target's complexities. Trainees are encouraged to refer back to the Phase 2 magnitudes in order to assist in the drawing of the Phase 3 sketch. Advanced viewers sometimes refer back to both Phase 1 and Phase 2 data.

To begin, the viewer obtains a new piece of paper, places the page number in the upper right-hand corner of the page, and writes "P3" centered at the top of the page. The paper is normally positioned lengthwise (the long side is horizontal). The viewer then begins to draw by quickly feeling around the page. The intuitions will suggest lines or curves at various positions. The beginning viewer is told not to edit out anything, but just to draw the lines as he or she feels them to be.

I once had a student who would simply not draw anything for the Phase 3 sketch. After I repeatedly encouraged him to sketch something, he finally looked at me and declared that he knew it could not be correct, but he could not get the idea out of his mind of a circle with what appeared to be many lines originating from the center of the circle and radiating outward. He then drew the sketch in order to show me what he meant. As it turned out, the sketch was a nearly perfect representation of the roof of a circular building that was the center of the target. The picture of the building that was being used to identify the target was taken from an elevated angle, and this viewer's sketch matched the angle and perspective exactly.

With Phase 3 sketches, the viewer need not understand what the sketch represents. As a general rule, it is impossible to know exactly what it represents. You can have an idea that there are people and a structure in the sketch, but you can never be certain. At best, you can only say that you feel there are lines here, curves there, and so on. Often simple drawings of people (i.e., subjects) or their ideograms are found in Phase 3 sketches. We never assume that such things really are subjects. At this point in the session, we know only that the drawings look like ideograms or sketches representing subjects.

After drawing any initial aspects of the sketch, viewers often run their hand or pen over the paper a couple of times (without actually contacting the paper). Doing so can give viewers a feel for where other aspects of the target are located. Viewers should quickly add these additional lines to the sketch. Beginning viewers are often seen moving their hands over the paper in clear patterns without ever drawing in these patterns. This is another editing-out problem. Many beginning viewers also move their hands in front of their faces, as if feeling a target. Novices nearly always fail to record these movements on paper, and have to be encouraged to do so. For example, if the target is a mountain, many students have been observed moving their hands in front of their faces tracing out the outlines of the steeply sloped mountain, even to the point of outlining the rounded or pointed peak of the mountain.

After finishing, students should look back at the dimensional magnitudes recorded at the end of Phase 2. Sometimes a glance at these magnitudes will trigger the sense of additional areas that need to be included in the drawing. For example, sometimes a student will write "tall" or "towering" as a vertical dimensional magnitude. Checking the Phase 3 sketch, the student may then perceive where this tall or towering thing is, and include it in the drawing.

In general, Phase 3 sketches are drawn rather quickly. Later, in Phase 5 (or in advanced versions of Phase 4), it is possible to draw meticulous and extended sketches. But the Phase 3 sketch normally has a sense of rapid data transference of initial impressions, not exacting drawings of the finer details. To spend too much time with details at this early point in the session would invite the conscious mind to begin interpreting the diagrammatic data. As an approximate rule, no more than 5 minutes should be spent on a Phase 3 sketch. A good Phase 3 sketch often takes less than a minute.

In Type 4 data situations, when the monitor knows the identity of the target, the monitor should interpret at least the basic aspects of the Phase 3 sketch immediately (while the session is still in progress). Listed here are a few useful interpretive guidelines.

- Perpendicular and parallel lines normally represent artificial structures or aspects of such structures.
- Wavy lines often suggest movement.
- People ideograms usually represent people.
- There is no way to estimate size with a Phase 3 sketch. For example, a circle could represent a golf ball or a planet.
- Some lines tend to represent land/water interfaces (where land and water meet, as on a coastline).
- Some lines tend to represent air/water or air/land interfaces.

Again, these interpretive guidelines are for the monitor's use during the session. Viewers should not try to use these guidelines to interpret a Phase 3 sketch on the spot. Viewers must concentrate only on recording the lines that represent or reflect the various aspects or parts of the target. After the session is completed, the viewer can spend as much time as needed interpreting the data in the sketches and elsewhere.

Chapter 7

Phase 4

THE MATRIX

Some of the most useful and descriptive remote-viewing information is obtained in Phase 4. It is impossible, however, to enter Phase 4 without first completing Phases 1, 2, and 3. Phase 4 works only after strong contact has been made with the target.

In Phase 4, remote viewers work with a data matrix. Each column of the matrix represents a certain type of data, and viewers probe these columns to obtain data. Phase 4 always begins with a new sheet of paper. The paper is positioned lengthwise. The viewer puts the page number in the upper right-hand corner and then writes "P4" centered at the top of the page.

The nine column identifiers of the Phase 4 matrix are written across the page from left to right. The first three columns represent data of the Phase 2 variety. The first represents data relating to the five senses of hearing, touch, sight, taste, and smell. This column is labeled with an S. The next column, labeled M, represents Phase 2 magnitudes. The third column is labeled VF, which represents viewer feelings.

The fourth column, not based on any of the earlier phases, is labeled E, which stands for "emotionals." Any emotions that the viewer perceives as originating from subjects at the target location are clearly emotionals. But the category can include much more. When intense emotions are experienced at a site, individuals commonly perceive these emotions even long after the fact. It is said that General Patton was able to feel intuitively the emotions of battle in an area even if the battle took place centuries earlier. Furthermore, some people feel "funny" about a site because of something that is to happen there in the future, not in the past. Thus, places vibrate with the emotions of events that have happened or will happen. In the slang of the day, certain places have "vibes."

For example, if a remote viewer is sent to the location of the Nazi concentration camp of Auschwitz at the current time, the viewer would normally perceive the buildings, the beds, the idea of a museum, and so on. But the viewer might also perceive the emotions of pain and suffering as relating to the site. Some viewers, depending on the flexibility allowed them, would be able to follow the emotions back in time to locate the origin of these feelings.

The emotionals column is placed next to the column for viewer feelings to help the viewers distinguish between these two types of emotionally related data. Viewer feelings are not the same as feelings perceived from a target, and the two should not be confused.

The next column describes physical things. These data can include perceptions of people, buildings, chairs, tables, water, sky, air, fog, planets, stars, vehicles, or anything else. The column for physical data is labeled P.

Some things are real but not physical. Remote viewers often perceive nonphysical things, such as beings, places, and so on. All of these nonphysical things exist in subspace. For example, a person without a physical body is real. Our souls are subspace entities, and when our physical bodies die we are no longer composite beings with physical and subspace aspects "glued" together. The subspace realm is at least as complex as physical reality. Basically, remote viewers have perceived that everything that exists in physical reality also exists—plus much more—in the subspace realm. Since remote viewers are using their subspace minds to collect data, it is natural that some of what is perceived will relate to the subspace realm. To differentiate clearly between physical data and subspace data, the subspace column is placed adjacent to the physicals column, and it is identified with the heading "Sub."

Novice remote viewers need practice viewing targets that have a large degree of subspace content or activity in order to become sensitive to subspace perceptions. This normally begins in the first week of training, but this exposure is continual, and improvements in perception follow a normal learning curve relating to how often they practice.

Data entered into the subspace column are exactly analogous to data entered into the physicals column. Subspace "things" are like physicals; they are just in subspace. If a viewer perceives other data that are subspace-related, but not "things," then the viewer places an S in the subspace column and then enters the data into the correct column at the same horizontal level as the S. This allows the analyst to differentiate between subspace and physical-related data entries that occur throughout the matrix. For example, emotions of subspace beings would be entered in the emotionals column, with an S being placed in the subspace column at the same horizontal level as these data.

The next column is for concepts, and it is labeled C. Concepts are intangible ideas that describe a target, but that do not relate to the five senses. All of the Phase 1 primitive and advanced descriptors are concepts, as are ideas such as good, bad, important, insignificant, inspiring, dangerous, safe, haven, work, play, fun, drudgery, adventurous, enlightening, attack, evolutionary, degraded, supported, healing, altruistic, evil, sinister, saintly, and so on.

The final two columns in the Phase 4 matrix correspond to two different types of deductions. The first is called a "guided deduction." A guided deduction is identical to a deduction except that the viewer actually probes the matrix in order to obtain the deduction. Reasons for doing this are explained in the following section on probing. The guided deduction column is labeled GD. The final column of the Phase 4 matrix is the deductions column, and it is labeled D.

To summarize, the Phase 4 matrix is:

S	M	VF	E	P	SUB	C	GD	D

Probing the Matrix

To probe the Phase 4 matrix, the viewer touches the tip of the pen in the appropriate column. Probing is delicate and should be performed with care. The pen should stay in contact with the paper for about a second. During that time the viewer perceives some information, usually—but not always—related to the column heading. If the pen's contact with the paper is too brief, then a sufficiently deep impression of the target will not have been made on the conscious mind. If the contact with the paper is too long, then the viewer risks having the conscious mind interfere.

After removing the pen from the paper, the viewer mentally searches for a word or brief phrase that describes the perceived information. This process is referred to as "decoding" the target perceptions. The viewer must decide on this word or phrase quickly, rarely more than three to five seconds after the probe. The viewer writes this description (usually one word) in the appropriate column.

Sometimes the viewer perceives a number of things when probing one column. When this happens, the viewer enters these data into the appropriate columns regardless of the column that was originally probed. For example, all emotional data go in the emotionals column, even if the emotional data are perceived when probing the physicals column.

When initially working the Phase 4 matrix, probing proceeds from left to right, skipping over the viewer feeling and deduction columns (explained in the next section). Viewers do, however, probe the guided deduction column. After probing a column, perceiving and writing something about the target, the viewer moves the pen down a bit before probing the next column. This results in a diagonal pattern of entries down the page. If a viewer perceives two or more pieces of related data, then the viewer places each of these in their appropriate columns at the same horizontal level, that is, without dropping down. For example, say a viewer perceives a brown structure. The word "structure" goes in the physicals column, and the word "brown" goes in the senses column, both at the same level.

Placing related data on the same level is essential for interpreting the data after the session is completed. If the viewer drops down a line after writing "brown" in the senses column and before writing "structure" in the physicals column, then the analyst would not know that it is the structure that is brown, perhaps concluding that something else at the target site is brown. Data can only be entered in a process that moves horizontally and down the page, never up. If the viewer at first only perceives a structure, then only the word "structure" would appear in the physicals column. However, if the viewer again perceives the same structure later in the session, but this time the color of the structure is also perceived, then the viewer again writes the word "structure" in the physicals column, but this time together with "brown" in the senses column at the same horizontal level.

Entering Viewer Feelings and Deductions

Viewer feelings are entered into the Phase 4 matrix only when they are felt. Viewer feelings are not data about the target; they are the subjective feelings of the viewer about the target. If undeclared, they will fester and contaminate the data still to be collected. Declaring them in the matrix removes their influence from the data flow.

Viewer feelings are entered into the viewer feeling column by first writing "VF—" followed by the feeling. For example, "VF—I feel happy," or "VF—This makes me sick." After declaring a viewer feeling, the viewer must put his or her pen down momentarily, as done in Phase 2.

Viewer feelings can happen at any point in Phase 4. Typically, viewer feelings manifest after probing either the emotionals or physicals columns. After a viewer feeling occurs and is recorded, the viewer returns to the point of last probing to continue the data-collection process.

Deductions are similar to viewer feelings in the sense that they can occur while probing any column. Whenever a deduction occurs, the viewer declares the deduction immediately by moving to the deductions column and writing "D-" followed by the deduction. As with a viewer feeling, the viewer should put the pen down while the deduction dissipates.

Guided deductions are exactly the same as deductions, except that they occur when probing the guided deductions column. While probing the matrix, the subspace mind knows that pressure is building in the conscious mind to attempt to deduce the identity of the target. Knowing this, the subspace mind can often ease the pressure by guiding the deduction out of the conscious mind at the correct time. By probing the guided deductions column, the viewer can rid the mind of the deduction at an early stage of its formation. This helps smooth the flow of the data and minimize the risk of having a developing and as yet undeclared deduction begin to influence the real data. One does not write "GD-" in front of the guided deduction, but does put the pen down after declaring it.

Remember that the subspace mind is still in control of the session when a guided deduction is declared. This is not the case with a normal deduction. With a deduction, the conscious mind interrupts the flow of data and inserts a conclusion relating to the meaning of the target or an aspect of the target. The subspace mind has lost control of the session at that point. With a guided deduction, the subspace mind does not lose control because it is "guiding" the removal of the deduction. Probing the guided deductions column allows this removal to be accomplished.

High- and Low-Level Data

One of the most crucial aspects of Phase 4 is differentiating between high- and low-level data. High-level data involve attempts to label or to identify aspects of a target. In the subspace realm of existence, information is not conveyed through words, but rather through direct knowledge gleaned from visual, sensory, conceptual, emotional, and other impressions. Indeed, this is the essence of telepathy—direct awareness of another's thoughts. Words are needed in the physical realm in order to convey meaning through speech or writing. If our words convey entire concepts, then we are describing something at a high level of identification. On the other hand, if we describe only the characteristics of what we perceive, we are working at a low level.

The difference is best shown through examples. If a target is an ocean shoreline, a remote viewer would likely perceive aspects of the target such as sand, the feeling of sand, wind, water, wetness, salty tastes, waves, the smell of lotions, and grass. These are all low-level descriptors of the target. High-level descriptors could be beach, ocean, shoreline, lakefront, tidal wave, and so on. The problem with high-level descriptors is that they are often only partially correct, whereas low-level descriptors are normally quite accurate.

The general rule in Phase 4 is to enter all or most high-level descriptors in the deductions column, reserving the data columns for low-level data. In the above example regarding the shoreline, an analyst studying the data would have no trouble identifying the low-level aspects as waves and possibly sand dunes. On the other hand, using the high-level data suggested above, the viewer could have been tempted to follow a story line created by the conscious mind of large waves, perhaps leading to a fabricated disaster scenario.

Entering high-level data in the Phase 4 matrix is very risky. Trainee viewers often want to obtain high-level data to demonstrate that they can identify the target. Yet novices should never try to obtain high-level data. You can describe nearly the entire universe using low-level data. In short, when we do remote viewing, we want to describe the target, not label or identify the target or its aspects. For example, if the target really is a tidal wave, then the viewer is safer describing a large wave, heavy winds, lots of energetics, destructive force, the concept of disaster, and so on. If the viewer thinks of a tidal wave, that idea can be entered as a deduction even though it exactly identifies the target.

To further clarify the difference between high- and low-level data, the following are some examples of each. In each case, it is safer deducting the high-level data while entering the low-level data elsewhere in the Phase 4 matrix. Maintaining a consistent stream of descriptive low-level data is perhaps the single most important criterion affecting the overall quality and usefulness of the session.

Low-level Data	High-level Data
explosive energy	bomb blast
sand, water, salty tastes, waves, perfume	beach
squirmy, primitive, scaly animal life	dinosaurs
tall structure with many floors	skyscraper
booming sound	explosion
sloping dry land with energetics or intense heat at top	volcano
many rooms side by side in multi-level structure	hotel
gathering of important subjects	U.N. Security Council

P4 ½

Most data that are entered in the Phase 4 matrix are single words placed in the appropriate columns. However, sometimes the remote viewer needs to say more than can fit in a column. This typically results after the viewer has recorded a number of low-level data items that he or she later feels to be connected in some way. A longer data entry that acts to organize or collect a number of separate gestalts is written as a P4 ½. This begins on the left side of the Phase 4 matrix. The viewer writes "P4 ½-" followed by a sentence or phrase, writing from left to right across the page. A P4 ½ entry is rarely more than one sentence, as this is to be avoided. It is better to write two or more P4 ½ entries sequentially than to attempt to write an extended discussion of the data. Entries that are too long risk shifting from recording perceptions to conscious-mind analysis.

Advanced remote viewers find P4 ½ entries most useful, especially after they have established thorough target contact. However, novices must watch out since they tend to use P4 ½ entries indiscriminately. Evidence of this is typically the appearance of a

P4 ½ entry that is not immediately preceded by a number of related single-word entries in the appropriate columns. Thus, the P4 ½ entries should ideally relate to and organize already perceived data, and they should definitely not appear to come "out of the blue."

P4 ½S

A P4 ½S is the same as a P4 ½, but it is a sketch rather than a verbal description. When the viewer perceives some visual data in Phase 4 that can be sketched, the viewer writes "P4 ½S" in either the physicals or the subspace column, depending on whether the sketch is to be of something in physical reality or subspace reality. The viewer then takes another piece of paper, positions it lengthwise, labels it P4 ½S centered at the top, and gives it a page number that is the same as the matrix page containing the column entry "P4 ½S," with an A appended to it. Thus, if the entry for the P4 ½S is located on page 9, then the P4 ½S sketch is located on page 9A.

THE "BIG THREE" AND "WORKING THE TARGET"

1. Probing the Matrix "Raw"

Probing the Phase 4 matrix has three distinct stages. When first entering Phase 4, the viewer simply probes the matrix as described earlier. This is referenced as probing the matrix "raw." Novices are instructed to obtain at least two pages of Phase 4 data, in order to prevent the viewers from giving up too easily. Beginning viewers are usually quite skeptical about their own data at first. Since this skepticism is rooted in the conscious mind, it is not a serious concern during training. Indeed, having the conscious mind preoccupied with skeptical thoughts can be a real advantage for a novice, since it clears the way for the subspace mind to slip the data past the reviewing processes of the conscious mind.

Working the Target

Advanced remote viewers treat their entry into Phase 4 as a means of obtaining crucially important information about a target. This requires them to continue longer in Phase 4 while they "work the target," the process of following a subspace signal intuitively through all of its leads. Viewers obtain a rich collection of data by "looking around," so to speak. If they find a structure, their intuitive sense tells whether it is important to know more about the structure. They describe it more thoroughly, moving inside the structure when needed to complete the description. The viewers describe the surface on which the structure is located. They may also describe the physical activities of the people outside and inside the structure, even locating a significant person who may be crucial to resolving the target cue. All of this is felt through strong intuitive tugs that direct the viewer's awareness in the appropriate directions.

Working the target also includes tying together low-level data in P4 ½ entries. When a viewer works a target, the viewer typically perceives some physical item and describes this item in low-level terms. This observation leads to another related observation, which in turn leads to another, and so on. After a sufficient number of low-level observations have been made, the viewer begins to "connect the dots," so to speak. A statement that pulls it all together, made as a P4 ½ entry, is itself a low-level description of the target or a fragment of the target. The statement does not label the target aspect.

For example, let us say that a viewer perceives wind, circular energy, extreme force, small flying pieces, and a vortex, all of these things being entered in the columns of the Phase 4 matrix. The viewer could then state the following P4 ½: "Windy circular energy in a powerful vortex containing lots of small flying pieces." The viewer could also declare a deduction of a tornado. The word "tornado" is high-level, since it clearly labels the phenomenon. The description in the P4 ½ entry remains low-level, even though it ties together other low-level data entries. The viewer then continues on to the next group of objects in a similar fashion. This is the classic method of working the target.

2. *Returning to the Emotionals*

After a while the flow of data will slow, and further working of the target becomes repetitive and unproductive. The viewer must then execute the second of the "Big Three" matrix processes. Even though the viewer has been regularly probing the emotionals with each horizontal pass through the Phase 4 matrix, a special trip back to the emotionals column often restarts the data flow. The reason is that the viewer's attention has been on various aspects of the target, and the emotionals data perceived earlier may have been related to those aspects, such as the sense of anger that resulted from an argument that took place within a structure. Returning specifically to the emotionals column for a special probing allows the subspace mind to shift its attention to other emotional data that could be more generally related to the target.

For example, let us say the remote-viewing target is the hostage crisis in Peru that began in December 1996. In this case, a group of Marxist guerillas attacked Japanese embassy facilities in Peru and held a large number of hostages until a Peruvian commando raid rescued nearly all of them in late April 1997. In the initial approach to the target, a viewer may perceive fear among the hostages as well as aggression among the guerillas. The viewer may describe two groups of people in a structure, with one group controlling another. After the data flow slows, the viewer returns to the emotionals column and probes it again. This time the viewer might perceive emotions of concern and concentration. This leads to perceiving the concepts of making a plan, waiting, rescue, high-level political involvement, and a commando operation. The viewer may also begin to perceive other people related to the target, such as a central figure (deducting a president), people with uniforms (deducting military personnel), and all this within a foreign setting (deducting Latin America). Note that the word "deduct" is used in the sense that it is a deduction being removed from the data flow.

Data for emotionals often lead to other physical and conceptual data. This is because the emotions of people at a target site tend to reflect what is happening around them, which in turn is grounded in their physical setting. Returning to the emotionals column also helps avoid what is known as the "doorknobbing" problem, in which the viewer focuses on one aspect of the target (such as a doorknob) while missing the broader picture (such as what else is going on in a room). Once the data flow is reinitiated, the viewer continues to work the target in the same manner as before.

3. Probing the Phase 3 Sketch

After restarting the data flow by returning to the emotionals column, the collection of data will eventually begin either to slow or to become repetitive as before. At this point the viewer returns to the earlier Phase 3 sketch and begins to probe various aspects of the sketch. Remember, when the viewer does the Phase 3 sketch, it is impossible to know exactly what it represents. However, it does represent the viewer's initial visual impression of the target, especially with regard to the arrangements of lines and shapes. By placing the point of the pen in various locations of the sketch—probing—the viewer is shifting the focal point of his or her awareness around the target location. This allows the viewer to reinitiate the flow of data once again, and the viewer returns to the Phase 4 matrix to enter the data in the appropriate columns.

When probing the Phase 3 sketch, the viewer is not trying to label or identify specific features of it, although these can be described in low-level terms. More generally, the viewer is simply using the sketch to obtain other low-level data by shifting his or her attention from one location to another. Viewers can probe lines in the Phase 3 sketch, resolving some of their meaning using the primitive and advanced descriptors of Phase 1. This is a good way of determining if there are structures or beings at the target site if this has not already been determined.

The viewers can also look for the following interfaces in a Phase 3 sketch: land/air, land/water, air/vacuum, land/vacuum, air/water. This is very helpful in determining various geographical features of the target site. For example, let us say that the viewer has determined that a structure at the target site is located on top of a flat surface. If the viewer probes below the structure and finds water, and then probes above the structure and finds air, the viewer then knows that the structure is floating on water and is probably a boat (which is a useful deduction). If the viewer determines that there is a structure in the Phase 3 sketch, and that the structure has air inside and vacuum above and below the structure, then the structure is most likely in space ("spacecraft" would be a deduction). If the structure is on a flat surface, and the surface is hard and natural (and thus land), and above the structure is air, then the viewer knows that the target involves a structure on flat land. If the viewer probes on both sides of a line in the Phase 3 sketch, finding water on one side and dry land on the other, the viewer knows that the target involves a land/water interface, and may deduct a beach.

CUING

The basic mechanics of cuing involve the viewer writing a word in an appropriate column (in either parentheses or brackets) and then touching the word with the pen. The word written in the column is the "cue." Using the pen to touch the word focuses the attention of the subspace mind on target aspects relevant to the cue. The resulting stream of data are then entered into the matrix in the appropriate columns below the cue.

Words that originate from the viewer's own data are entered in the appropriate column in parentheses ( ). Cues originating from a monitor, or not from a viewer's own data, are entered in square brackets [ ]. If the monitor's word(s) are used to construct a cue, then the cue should be non-leading and closely tied to the viewer's existing data. For example, if a viewer perceives a building, the monitor may suggest that the viewer cue on "activity" by writing the word in square brackets in the concepts column, then probing the word and entering the resulting data in the appropriate columns of the matrix.

MOVEMENT EXERCISES

There are three types (called "levels") of movement exercises. All levels can be performed after spending some time in Phase 4.

Level One

These exercises essentially return the viewer to a modified form of Phase 1. An ideogram is drawn and decoded, and the person returns to Phases 2 and 3 before arriving again at Phase 4. This is done for one of two reasons. If the monitor is concerned that the viewer may have wandered off target, a level-one movement exercise nearly always returns the viewer to the target. The other reason is that the viewer may need to relocate to another area related to the target that may be substantially different from the area being probed so far. The new Phase 1 through Phase 3 information may help the viewer differentiate between the two target-related sites.

These cues are written from left to right across a Phase 4 matrix. Usually a half page is needed; otherwise, a new piece of paper is used. The Phase 4 matrix does not need to be rewritten on the new paper, but do include the page number. Immediately after the viewer writes the cue, the viewer places the point of the pen to the right of the cue and draws an ideogram. The ideogram is then decoded in the manner of all Phase 1 ideograms. Only one ideogram is used in a level-one movement exercise before moving to Phase 2. The following is a list of cues used for level-one movement exercises, beginning with the most common:

1. "From the center of the target (or target site, target area), something should be perceivable." Most level-one movement exercises use this cue, especially for the first such exercise.
2. "From 1,000 feet (or an alternative lengthy distance) above (or to the north, south, east, or west) of the target, something should be perceivable." This cue should be used only if it is unclear where the viewer is relative to the surrounding (viewed) environment. This cue should only rarely be the first level-one movement exercise since it essentially moves the viewer away from the center of the target, which is usually the most important part of the target.
3. "Immediately to the left (or right, in front of, behind) the target, something should be perceivable."

4. "From the center of the target area (or site), the target person (or object) should be perceivable."
5. "From inside the structure, something should be perceivable."

Level Two

Level-two movement exercises are used to move the viewer from one location or target-related item to another without the viewer having to leave Phase 4. This exercise is not such a total break as a level-one movement exercise, but neither is its shift in focus as subtle as a level-three exercise. The cue is essentially the same regardless of the situation, with only locational words being changed. Here is the cue:

"Move to the [new target location or item] and describe."

In this cue the "new target location or item" should originate from the viewer's own data. The monitor normally does not insert his or her own words here, except to focus the viewer's attention on some particular generic component of the target. For example, the "new target location or item" can include phrases such as "target subject," "target subjects," "target object," and so on.

The level-two cue is written across the body of the Phase 4 matrix, from left to right. The viewer then continues to enter data in the same matrix in the normal fashion after writing the movement exercise cue. There is no ideogram in this exercise. However, I personally find it useful from time to time to probe the last letter of the word "describe" in the level-two cue in order to refocus my attention.

A level-two movement exercise can be temporal as well. This exercise cue follows the following format:

"Move to the time (or period) of [temporal identifier here] and describe."

In this cue, the temporal identifier must be clearly connected to the viewer's earlier data. For example, if the target is a pyramid in Egypt and the viewer describes a pyramid structure, the monitor could give the cue: "Move to the period of construction for the structure and describe."

Level Three

This is the most subtle of the three movement exercises. It shifts the viewer's awareness without breaking the previous flow of data. The movement is executed by placing a very brief cue (usually only one or two words) in the appropriate column of the Phase 4 matrix and then having the viewer touch the cue with the pen and begin entering data. The cue can be a word originating from the viewer, entered using parentheses ( ). If the cue originates from the monitor, square brackets [ ] are used. Cues originating from the monitor should be used only rarely in Phase 4, and if used, should be of the most generic variety.

For example, the viewer perceives two beings—a male and a female—separated by, say, a road. The viewer could move from the male to the female by putting "(female)" in the physicals column, probing this with the pen, and then continuing with the collection of data in the Phase 4 matrix.

One particularly interesting level-three movement exercise is a deep mind probe. In this the viewer enters the mind of a person in order to obtain thoughts and personal character information. There is an ethical component to this exercise, though. The subspace mind of any person being remote viewed will be aware of this activity even if the person's conscious mind is not. This is yet another reason why I recommend that all remote viewers meditate regularly in order to remove as much of their own stresses as possible before entering the mind of someone else. It is mandatory to do no harm while remote viewing.

A deep mind probe is performed by writing "[target person]" in the physicals column and "[deep mind probe]" in the concepts column. The viewer then touches each of the words in each phrase once with the pen, and enters the relevant data in the matrix, usually in the emotionals and concepts columns.

A level-three temporal movement exercise can be obtained by using event- or action-related cue words. These cues need to be clearly connected to the viewer's data. Such cues are entered in square brackets [ ] in the concepts column in the Phase 4 matrix. In introductory and intermediate remote viewing courses, "activity" is normally the most frequently used temporal level-three cue.

There are three other chapters included as Appendices 1, 2, and 3 containing technical material for advanced SRV procedures. The first chapter deals with specialized procedures that are used in Phase 5 of Basic SRV. The second chapter explains Enhanced SRV, which is a highly interactive and flexible form of the protocols, while the third chapter describes procedures used to analyze societies.

PART II

Verifiable Targets

Chapter 8

Death on Mount Everest

Climbing Mt. Everest is one of the most dangerous adventures known to man. Many have died in the climb, and it is not unusual for even veteran climbers to encounter a life-threatening situation, such as a sudden storm, or just bad luck. When death approaches a climber on those steep slopes, there are nearly always moments for reflection, time to ponder the wisdom of one's actions. Within the consciousness of those moments resides a telling truth, a blunt realism. To witness this realism, this deep awareness of what one has done without the illusion of grandeur to cloud the vision, is to witness a moment of consciousness in a state of great purity. When some die, they may feel alone, with no one to share their last moments of wakefulness. Yet situations do occur in which those who die unknowingly share these moments with a witness of the soul.

15 May 1997
3:50 p.m.
Atlanta, Georgia
Protocols: Basic SRV
Target coordinates: 6861/2306

My first ideogram feels hard and man-made. Yet there is

something unusual about it, and I declare that the ideogram represents land. My second ideogram is again hard and man-made, but it is a shape that is typical for a structure. The third ideogram is similar to the second, but it feels hard and natural. I deduct the idea of a mountain.

I perceive airy sounds, like that of wind. The textures at the target site are rough and rocky. The temperatures are cold, and I deduct Mt. Everest. There are colors of blue, white, brown, and tan. The level of luminescence is bright with high contrasts. I taste something salty, like sweat, and again deduct Mt. Everest. The air has the smell of ozone. The air is thin and smells fresh. The magnitude of the dimensions are tall and towering verticals, wide and expansive horizontals, long sloping diagonals, and heavy mass. There is a moderate level of energetics at the target site. My Phase 3 sketch resembles a mountain covered with something like snow at the top. I deduct a volcano on my sketch.

In Phase 4 I perceive bright white and blue colors, and I deduct snow. The contrasts are very high. Again, the textures seem rough and rocky. The magnitudes are clearly tall and towering, and the air seems thin. I am detecting the mental flavor of consciousness at the target site. There are subjects, and I deduct "climbing a mountain." There are rocks, a path, and I sense the concept of passage. Within the subspace arena, I detect interest. This target is associated with the concept of achievement. "Climbing Mt. Everest" emerges as a guided deduction.

Bright white light is everywhere, as is the sense of something tall and towering. The concept of achievement is mixed with the

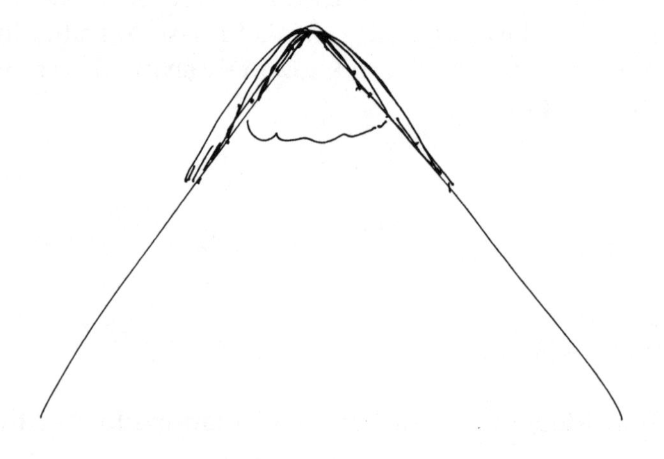

emotions of thrill and tension. The subjects are wearing rough clothing, like jeans. The clothes feel like they are sturdy and rough, the type needed for rugged outdoor activity or work. Now I perceive that the subjects are male. There is a group involved with an expedition. There is a male leader in the group. I deduct the ideas of hiking, backpacking, and a mountain. I again perceive clothing, rocks, a narrow path, and trees. I do not perceive a structure in my current position at the target.

This target is associated with a variety of overlapping concepts. Here there is achievement mixed with the ideas of striving and overcoming hardship and difficulty. I have a guided deduction of mountain climbing.

Focusing on the target subjects, I perceive that they are trying to concentrate, to keep their minds focused. They are having difficulty thinking, the way someone would if there was not enough oxygen to breathe. Their minds wander, and they forget. Conquering the problems with their minds is a challenge to them. They are working to achieve a goal. I feel subspace emotional energy that is supportive of these activities, but not in the essential "this must happen" sense. The bright light of the sun, the tall towering magnitudes, the difficulty in concentrating, the determination of the subjects, all these are mixed together.

I execute a collective deep mind probe on the target subjects. These people need to do something for their own sake. This is very important to them. It is not a life-and-death issue, but it feels like that in their own minds. They seem to have made it a "do-or-die" situation.

While the emotions of the target group contain both tension and fear, there is also suppressed exhilaration. The group is worried about shelter and supplies for some project. I deduct pickaxes, small shovels, and a tent. I also deduct that the group is on a journey.

Returning to my Phase 3 sketch, I find the air thin and cold. Indeed, it is freezing here. There is human occupation far below, and I deduct farms and villages. I also perceive that something is hot at the target site. Cold is everywhere at the top of the mountain, but hot is localized somewhere else. I conclude that the target appears to contain a tall mountain that is both hot and cold at the top. Yet it is much more cold than it is hot. I again note that the air is thin.

Moving to the precise center and time of the target, I sense that the subjects are going up, or climbing over. They feel exhilarated, but there is also grave danger in the activity of climbing this mountain. I get the overwhelming sense that their single-minded focus is "going up, going up, going up and over."

Discussion

After I finished this session, I was told the target cue, "Mid-May expedition up Mt. Everest in which two guides and five hikers were killed (circa May 1997)." This is an unusually clear remote-viewing session. Even the most advanced viewers often have significant areas of ambiguity mixed with some decoding errors in their work. I was quite surprised with the accuracy of the physical target descriptions in this instance. Research on the target also indicated that much of my description of the emotional dynamics among the group members was particularly insightful. Following the completion of this session, I briefly entertained the notion that my capabilities might have catapulted upward permanently, and that all of my future sessions would have this level of profound target contact. Alas, I am just as human as everyone else, and this level of clarity is still more the exception than the rule.

But readers should examine this session carefully. All that I was given before beginning this session were the target coordinates, chosen by my tasker from a table of random numbers. I had no monitor to lead me. Nor had I ever heard or read about this expedition before doing this session.

There is no way to associate a probability of psi functioning with a session that contains such a high level of profound target contact. Skeptics of remote viewing may insist on constructing some imaginary rationale to explain this session without addressing the real causal link involving consciousness. My soul witnessed this event, and it was through my conscious recognition of my soul's perceptions that I was able to write down these descriptions.

Chapter 9

The Hubble Telescope Repairs

The following session is an excellent illustration of how the conscious mind and the soul interact. While remote viewing a verifiable space operation, my conscious mind tried to interpret the data in a fashion that allowed a close parallel to the actual raw data. Nothing is more important in understanding remote-viewing data than the fact that each individual has two separate minds that can function independently to a surprising extent. In this session, it is clear what information my subspace mind was trying to convey, and (by reading the deductions) it is equally clear how this information was interpreted (falsely) by my conscious mind. Such decoding difficulties can occur with any viewer, regardless of expertise, and a trained analyst is normally alert to strip conscious mind interpretations from the data.

Readers should examine this chapter closely to learn why it is so difficult to rely on what the viewer thinks he or she has perceived. Unless one is quite skilled in these matters, correct intuitions can be easily sidetracked by incorrect interpretations.

30 May 1997
10:56 a.m.
Atlanta, Georgia
Protocols: Basic SRV, Type 3
Target coordinates: 8810/4131

Employing Basic SRV, I observe in Phase 1 that the target involves both a structure and movement. I hear rushing or roaring sounds, like those of a jet. The textures are soft and fabric-like. The temperatures range from hot to cold. There are many primary colors, including blue and red. I deduct a balloon. The luminescence is incandescent and bright, while the contrasts are moderate. I perceive the taste of food, and I smell food as well. The magnitude of the dimensions include tall and high verticals, medium-width horizontals, curving diagonals with a round topology, and hot, fiery energy. My Phase 3 sketch is of a circular structure connected to a more rectangular structure. Due to the shape of this sketch, I deduct a hot air balloon.

In Phase 4, I perceive hot, fiery energetics. Something is round, and it is flying very high. I perceive cloth textures. Whatever is flying high feels empty or hollow in some way. Energetics are associated with this target. There are subjects who are very excited at the current time. I sense only a few subjects, most or all male. The primary target structure is empty or hollow. It feels hotter inside and cooler outside. The structure appears to be round or curved in shape.

Using my hands to follow the shape of the structure, I note

that it seems to "give" inward when pressed from the outside, reinforcing the idea that fabric of some sort is on the exterior of the structure. This structure also feels like it is the primary target aspect. There is the concept of thrill associated with this target. I again perceive woven textures, like cloth, on the exterior of the target. The subjects are very excited, and I am deducting the concepts of patriotism and the flag. I also deduct a hot air ride over Africa or Missouri. The structure appears to be near some surface, perhaps land. It may also be that some surface or horizon is visible from the perspective of the structure's location. The structure itself feels light in some way.

I observe that the subjects are wearing distinct clothing that seems to be intended for special warmth. They are flying, and it is a thrilling adventure for them. There appear to be approximately four subjects, and I am beginning to feel a female energy from one of the subjects. The other subjects clearly seem male.

I move my perspective into the structure. It has a complicated or sophisticated design. There seem to be strings or thin connectors inside. Otherwise, the structure appears light and empty or hollow. The structure is traveling, and the movement appears slow from my perspective.

Shifting my perspective to the target subjects, I execute a collective deep mind probe. The subjects have a mixture of emotions, including tension, combined with others of varying degree, depending on the individual. I again sense three males and one female. They appear to be standing together, looking outward. They are participating in some kind of activity that involves manipulating things. I again sense that a hot flame is near the target.

I re-orient myself to the center of the target with a movement exercise. I perceive that the structure is round, in the sense that it has a curved topology. I again note the excitement of the target subjects, and I feel their sense of adventure. I draw a sketch of the target structure. The sketch resembles a flying hot air balloon. The horizon is visible below the structure. The target subjects appear to be near one side, or perhaps the bottom of the structure. The hollow structure seems to be above them.

The structure itself has thin walls. There seem to be higher temperatures inside the structure than outside the structure. But the outside does not feel frigidly cold. The subjects are in an area near the lower part of the structure. This area is compact, dense, and congested. There are many technological devices in this area. Below the structure I perceive empty space, and when I probe the apparent horizon, I perceive solid land.

Discussion

The target cue for this session is "The shuttle *Discovery* mission to repair the Hubble Space Telescope (mid-February 1997)." If one ignores the deductions (which is the appropriate thing to do since deductions are analytical conclusions, not raw data), this session is quite accurate. I perceived a flying structure, a small group of people working on that structure using technological devices, and the horizon of the land below the structure.

When I first analyzed the session, I was struck by how strongly my mind wanted to interpret the data for the structure as representing a balloon. In particular, I was concerned about the perceptions of cloth or textiles surrounding the structure, which was the aspect that was leaning me in the balloon direction. I could not see how this could possibly be true, since the Hubble Space Telescope is certainly constructed of metal. I then conducted some research for the target event using the CNN website. As it turns out, I learned that the telescope is covered with reflective insulating fabric that is used to protect the exterior from solar radiation. According to the reports, the shuttle crew were concerned about the extensive damage that they observed to this protecting covering (due to unexpectedly high wear and tear). Since there is a layer of insulation between the fabric and the telescope, the exterior "gives" when pressed.

I could obtain little information about the crew of the *Discovery*. The CNN reports made no mention of a woman, or the number of total crew members. Thus, these data may have contained some inconsistencies. Nonetheless, I did accurately perceive that the target subjects were few in number.

By this time readers should be able to clearly identify the tension that exists between the conscious mind and the subspace mind with regard to the exchange of information. The subspace mind perceives raw information that is not processed logically or verbally. The conscious mind must use words to describe the intuitive content of this information, including sketches of the low-resolution images. The conscious mind and the subspace mind do not approach awareness from the same point of view. There is a translation problem that is compounded by the fact that the essential natures of the two minds are different.

All this implies that remote viewing is a skill that takes considerable practice. A person does not learn the procedures and instantly perceive accurately across time and space. Rather, accuracy increases with practice. Just as all artists must practice regularly in order to perfect their skills, remote viewers need to view often in order to maintain and improve their skills.

Chapter 10

Centers for Disease Control

I used to live near the Centers for Disease Control (CDC) in Atlanta, Georgia. Every day I drove past the oddly assembled buildings that make up that complex. Historically, buildings were added as needed and as funding allowed. The architects always changed, and no common theme developed to guide the evolving physical appearance of the site. What exists now is a hodgepodge of structures with differing shapes, sizes, and designs. Some roofs are flat, while others are slanted. Smoke stacks emerge from one building while a dish antenna tops another. A large sloping parking lot is located behind the complex. From a design perspective, this facility is a mess. From a remote-viewing perspective, its complexity makes it a challenging target.

Is this then a chapter about remote viewing a collection of buildings? No. Readers should remember that the most complete collection of deadly viruses and bacteria known to man are housed in this facility. Now, put yourself in the position of a strategist trying to locate the biological warfare weapons of a terrorist organization. Remote viewers could be assigned a target cue that would direct them to some suspect facility.

Depending on the artistic capabilities of the viewer, the data and sketches may be of a type reminiscent of those contained in this session. The information would hopefully not only describe aspects of the structures at the site, but the behavior and appearance of the personnel working in the facility as well.

The analysts would then compare the data with known facilities in the area under investigation. If a match is suspected, other forms of intelligence would be used to obtain corroborating or confirming evidence. This is the way remote-viewing data can be used to initiate leads that are later followed up. The same would be applicable not only to search operations relating to biological warfare agents, but also to other situations in which physical descriptions of buildings and their associated personnel are required. In this day and age when weapons of mass destruction can be obtained by terrorist forces, sessions resembling that presented here may be common in our future.

12 August 1997
12:37 p.m.
Atlanta, Georgia
Protocols: Enhanced SRV, Type 3
Target coordinates: 2680/1114

My first ideogram feels soft and artificial. It represents something that contains the colors red, blue, and white. I perceive a woven texture, and I hear a snapping sound. My sketch is of two flags, one of which is flapping in the wind.

The second ideogram feels semi-hard. I again hear sounds of wind, and the smells are of outdoors. Something is heavy, tall, and towering at the target. My sketch is of a tall rectangular structure. The third ideogram feels hard and man-made. It also feels heavy and dense. The colors are tan and light brown. The sketch is of a circle.

The fourth ideogram feels hard and man-made, and I perceive that it represents a structure. The colors are light brown and gray. The textures are a mixture of rough and polished. The magnitudes of the dimensions are thin and short, narrow and compact. The sketch is of a short rectangular structure.

My final ideogram also feels hard and man-made, representing a structure. I again perceive the combination of rough and polished textures, as well as the colors of gray and green. The sketch of the structure suggests that it has a steeply slanted, curved roof.

My Phase 3 sketch suggests that the target involves a rectangular structure in the background with a flag on a flagpole in the foreground. I begin Phase 4 observing that the target is irregularly shaped. It is gray, bright, and polished. There are subjects in the irregularly shaped target structure. The structure is complicated to draw with detail. Currently, it appears to be hit by bright light. I observe the highly reflective surfaces of the target structure. There is technology associated with this target. It is both big and tall, and its topology is both curving and straight.

There are multiple subjects at the target. They are wearing clothes that seem like smocks or lab coats. Their long white clothing hangs down over their bodies. There are both males and females at work, and I sense a mission in their activity.

I sketch one aspect of the target structure. It has a single, curved, steeply tilted roof. Its walls are reflective, at least partially. Some of the target subjects within the structure wear uniforms. I execute a collective deep mind probe of the target subjects and perceive the emotionality of worry, focus, and concentration. I sketch one of the target subjects. This subject is a male, and he is wearing a long white lab coat. The other target subjects are focused on work activity. I note that the floor on which they walk is polished and shiny. I draw another sketch of two subjects wearing lab coats within the target structure.

To refocus myself at the center of the target, I execute a movement exercise. After returning to Phase 4, I observe technology that is compact. The target subjects are still focusing their concentration in a work environment. Their white clothes still fully cover their bodies loosely. I observe their shoes, their hair, and their faces.

I now note that something is circular at the target, and I sketch a round structure with lines that radiate outward from the center. It looks like spokes on a wheel. There is technology here. One facet of this technology is associated with the transmission of radio signals. Looking around the target site, I observe that the target structure is on dry land. I draw another aspect of the target structure, which is short and rectangular. Overall, my impression of the target structure is that its topology is complex and irregularly shaped.

Discussion

The target cue for this session is "CDC (current time)." As mentioned earlier, the architecture of the CDC is highly complex. It is not one structure, but many differently shaped structures grouped together. There are flags on flagpoles in front of the main entrance to the complex, exactly as described in this session.

Note that I perceived the activities and appearance of the personnel who work at the CDC. Long white lab coats are virtually a uniform among the researchers and technicians. Also, they indeed have a mission. CDC exists to serve humanity by assisting in the control of infectious diseases. There are also uniformed guards throughout this facility. (Remember that every deadly infectious agent known to man is stored in these buildings.)

If I were a human living on a far-off planet, it would be possible for me to remote view this facility by targeting the central location where efforts to control worldwide diseases are coordinated. This session, combined with related sessions from other viewers, would enable the analysts on that distant world to develop a description of the facility as well as the personnel that work there. In the absence of transportation capabilities, this description would be labeled unverifiable. But because we can walk up to this target and touch it, the current session seems more real. Were we to know of the remote-viewing efforts of this distant civilization, we might ridicule their hesitancy to accept as real the information that they obtained using methods that work quite well with other targets that are verifiable to them.

This, of course, is the current situation in which we find ourselves as a species. We note with interest the accuracy of a session in which a verifiable target is clearly described. But we balk when we are asked to seriously consider the results of esoteric targets, regardless of how many times the experiments are repeated, and regardless of the controls used.

Chapter 11

THE KU KLUX KLAN

Ever wonder what really happens at a reunion? This chapter presents remote-viewing data of a recent reunion of the Ku Klux Klan in Georgia. From these data it is clear that Klan members share a strongly held perspective that they hold in memories, folklore, and stories. But their reunion also invokes a heritage that connects them with a living history. I theorize that when groups form on the basis of strongly felt ideas that are connected to earlier events and conditions, the consciousness of the people who shared those ideas in the past reverberate in the present.

Time is no barrier to remote viewing. Similarly, even though the individuals concerned may not be consciously aware of this, time is totally transparent to the souls of those beings who share a common viewpoint. This relationship exists regardless of whether or not two groups of such beings are separated by decades, centuries, or longer.

27 May 1997
2:05 p.m.
Atlanta, Georgia
Protocols: Basic SRV, Type 3
Target coordinates: 1443/0210

My first ideogram indicates that the target involves a mountain. Other ideograms appear to address the land or environment around the mountain. In Phase 2, I hear something airy, and I perceive the crackling sounds of fire. The textures at the site are sharp, smooth, polished, and possibly painted. Temperatures are both cold and hot. There are blue and white colors with moderate luminescence. I taste blood and salt. The air smells cold and fresh. The magnitudes of the target dimensions include something tall, towering, and steep. The horizontals are narrow while the diagonals are long and sloping. Something is both heavy and open at the target site, and I perceive energetics of some type. My Phase 3 sketch is of a mountain with a curved base.

In Phase 4 I immediately deduct both a mountain and Mt. Everest. The target feels very open in some way. I smell something burning. One aspect of the target feels heavy, long, and sloping, and I deduct a pyramid. I also begin to perceive the emotions and thoughts of subjects at the target site. There is a stone structure that feels man-made. This stone structure is large, and I again deduct a pyramid. The textures are sandy while the colors are tan and brown.

There is another structure on land at the target. This structure is made of wood. I can perceive the walls as well as the interior and exterior of the structure. The structure has doors and other openings, and it is made of natural materials. Something about this target feels old, even ancient.

There are thoughts and emotions associated with the target site. But I am not perceiving the subjects who had these thoughts as currently present. There are other subjects present at the target site, and they are fewer in number. They are near or in the wooden structure. There is furniture in the structure, and a hallway with rooms off to the side. I deduct the rounded mountain, Santa Fe Baldy. Focusing again on the emotions at the target site, I perceive a few subjects physically, but the emotions of many.

I execute a collective deep mind probe of the target subjects and find their minds to contain the concepts of being underground or hidden. I deduct the idea of hiding. Drawing another sketch of the target, I place the structure with the subjects in the foreground, and the mountain in the background. There are lots of emotions associated with this target. They are not bad emotions, just many of them.

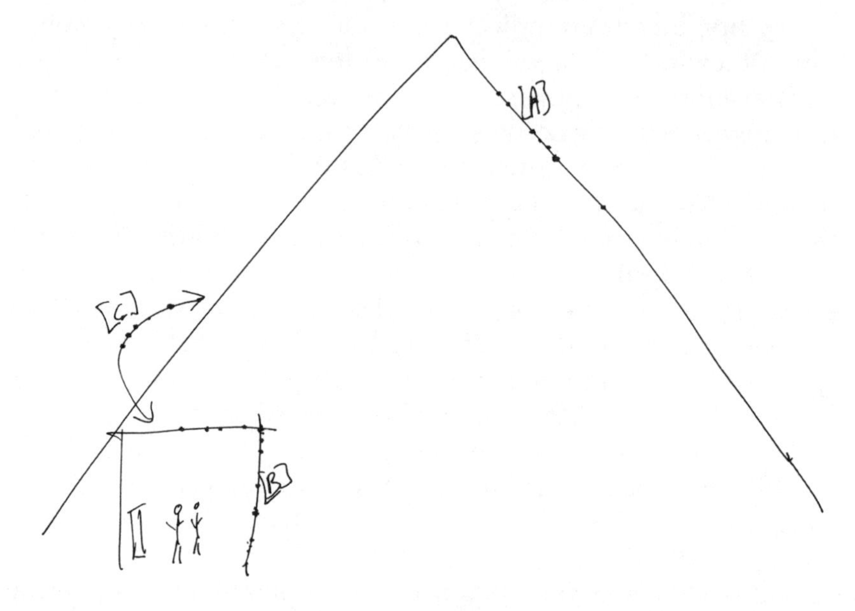

This target is connecting to disparate things, places, times, events. There is history here. Something old has passed away. The mountain is of some significance to the target subjects. It is an object of study, examination, and interest. It represents something to these subjects, something ancient, something historical. Whatever it represents generates many thoughts and emotions in their minds. Oddly, I still perceive only a few physical subjects, but the emotions of many crowd the collective consciousness associated with this target.

Discussion

The target cue for this session is "1996 annual Ku Klux Klan meeting / Stone Mountain, Georgia / event (20 July 1996)." This reunion of Ku Klux Klan members took place near the base of Stone Mountain. This is a large, rounded granite rock that protrudes from flat land. On one of its faces is carved a huge sculpture of Confederate heroes in the Civil War. This explains why I perceived the mountain to be man-made, and why so many deductions appeared comparing it to a pyramid. This also explains the deduction of Santa Fe Baldy, another round-top mountain.

The most interesting aspect of this session is the sense of connectedness to the past that I perceived from the target subjects.

These beings felt a nostalgic longing for days gone by. But most interesting, my subspace mind clearly perceived the emotions of many others who were not there physically. This was not just a reunion of physical beings; this was a reunion of the spirit.

There seem to be two possibilities to explain this phenomenon of spiritual connectedness across time in this instance. First, perhaps the racist philosophy of the target subjects, as well as their emotional attachment to history, induced many of the souls of those who are no longer physical to identify with the modern-day Klansmen. The Klan of the past was with the Klan of the present at the reunion, literally, in spirit. Alternatively, perhaps the minds of the target subjects used the vehicle of nostalgia to transcend time, to make a direct connection to the days of burning crosses and lynchings. Remember, the subspace minds of these people are as transparent to time as my own. Possibly I perceived the emotions of many because I followed the awareness of the target subjects backward through time. I do not know which of these two theories is more probable.

Chapter 12

North Korean Society

In 1997, I developed and tested new remote-viewing protocols that could be used for sociological and political analysis. SRV is used to observe physical things. But something entirely different was needed to remote view a society and its government. I was not interested in merely describing the buildings that house a government, but the government itself. For example, I wanted to be able to use remote viewing to perceive whether a society is democratic or authoritarian, the relationship between a populace and its leadership, the organization of groups within the society, and so on. The new protocols that I developed are called the SRV Social and Political Protocols (or SPP), and they are described in Appendix 3.

This chapter presents a remote-viewing session in which I use the Social and Political Protocols to describe a known human society. I performed this session completely blind to the nature of the target (Type 3 data). The target is verifiable, and I conducted some blind analysis after the session was completed during which I correctly identified the target from a list of five possible targets, four of which were decoys. This chapter (as well as some subsequent chapters) uses a few terms which are not fully explained until Appendix 3. Some readers may want to read this appendix at this time. However, many readers should find most of this chapter accessible regardless. If there is any difficulty understanding a term or a phrase, you might wish to look up that item in Appendix 3.

11 December 1997
2:53 p.m.
Atlanta, Georgia
Protocols: SPP, Type 3
Target coordinates: 2468/4952

In Phase I, the first ideogram addresses the macro target, which is the wide-angle perspective of the relevant society. The ideogram describes physical beings, and a culture that contains subdivisions.

The second ideogram identifies a sub-macro culture. This smaller group is hierarchical in its structure, tightly knit in its internal organization, and highly focused, even single-minded, in its approach to its activities.

The third ideogram identifies an additional sub-macro culture. But this group is loosely organized with a common cultural theme. It is a diffuse culture with a central authority. Common ties between group members are what holds this culture together. Thus, the dominant glue of the society is cultural enforcement, not bureaucratic or police enforcement.

The fourth ideogram again identifies the macro target. I perceive the larger group and the fragmentation within that group. The groups within the larger target macro are loosely organized, with the apparent involvement of a governmental bureaucracy to maintain the organization.

Overview of the Target Macro

From within Phase 2TM, I perceive that the target has a large population size with only one physical genetic strain. There is a single concentration of authority of medium strength within the society. The quality of that authority is both diffuse and multilayered. The culture itself is moderately diverse, yet with a common theme.

From a political perspective, there is one dominant ideology that is widespread throughout the society. Interestingly, however, the strength of this ideology is weak. The political orientation of the society is similar or uniform throughout. There is a significant amount of group fragmentation as well.

The term institutionalization refers to the level of habitual behavior among the members of a society. For example, when people vote for the same political party in every election, it is said that their level of institutionalization is high. For this target, the level of overall institutionalization of behavior is regular, yet weak. While this indicates that people behave the same way over time, suggesting a lack of behavioral diversity within the culture, the potential exists for this highly patterned behavior to change in the future, possibly dramatically. On the other hand, the institutionalization of behavior for the dominant group in the society is highly rigid and very regular. Thus, the dominant group is different in this regard than the general populace.

This society is predominantly focused on the physical rather than subspace level. The psychology of the subspace side is detached from that of the physical side, which implies that the physical side has lost any real understanding of its subspace counterpart.

The macro-society tries to strongly control the sub-macro groups. The sub-macro groups resist that control to some extent, as they attempt to stay at arm's length from the coercion of the macro-society.

The significant leader of the macro-society is male. Conducting a deep mind probe of this leader, I perceive that the ideas dominant in this person's mind are those of control, persuasion, and manipulation. This person plays groups against one another in an effort to maintain control. He generates confusion and divisions using strategies that are overlapping. This is the old strategy of divide and conquer. The leader maintains control through group separation, not homogeneous cooperation. He interacts with various interest groups and does not allow any one group to dominate.

I then execute a consciousness map for the target macro. Again, this procedure allows a viewer to perceive basic aspects of a larger group's collective psychology. On the physical side, the rules of enforced social organization dominate the psychology of the group. Interactions between individuals and groups follow a set pattern aimed at eliminating deviant behavior. On the subspace level, the dominant emotions are those of excitement combined with angst. Indeed, there is somewhat of a competition among subspace beings to participate in this culture (by being born into it). For these subspace beings, this culture is a valued point of soul entry into the physical world.

To summarize the Phase 2TM results, the society feels diverse and loosely organized, following patterned behavior that is the primary factor in maintaining social coherence and conformity. It is not the police that maintain social adherence to the macro-organization, but a history of patterned behavior, and perhaps a coherent ideology that is supportive of that macro-behavior. In Phase 3TM, I draw a schematic representation of the society that includes a central dominant group (G1) and four peripheral groups (G2–G5).

G1: The Dominant Group

I began Phase 4GB by focusing on the dominant group (G1). The population size of this group is the largest relative to all other groups. There is only one genetic strain. The concentration of authority within this group is high, with a uniform authority distribution. The culture is very homogeneous. The ideology, or theme of ideas that dominates and differentiates the society, is historical and political. It is both widespread and strong. There is little or no group fragmentation within the dominant group. The institutionalization of behavior within this dominant group is high and, indeed, this is a defining characteristic. The level of this institutionalization of behavior is much higher than that of the larger society's average.

The activity of this dominant group is almost entirely focused on the physical level. Its use of telepathy is very limited. The subspace psychology of this group desires a relationship with the physical side. But the physical mentality has totally forgotten virtually the entire subspace realm.

The collective psychological relationship between this dominant group and the larger society is one of control. This group is slow to change. The larger society recognizes this group's dominance and offers no effective challenge to the status quo.

The significant leader of this dominant group is a political leader who recognizes the importance of the sub-macro groups more than is commonly recognized by other members of the dominant group. This leader does not operate without constraints. He does not have total power, and he can be effectively challenged. He needs to use all of his formidable skills of manipulation in order to maintain control of the dominant group.

Shifting to a typical non-leader member of the dominant group, I perceive from the mind of this person a sense of weakness and lack of control. The idea is an internal sense that the larger society is out of control. This individual has cultural ties to the larger group, and a similar ethnicity. This is not a tight bonding, but there is no ambiguity about membership in the larger group either.

From a consciousness map of this dominant group, I perceive a sense of patriotism that is both superficial and of mild strength. There is organization for the sake of organization within the society. There is no vision of contemporary significance that guides the society. From the subspace side, this dominant group has some kind of preferred position. There are subspace individuals that are active in supporting or being involved with this group for selfish reasons.

G2: A Peripheral Group

Continuing the group breakdown in Phase 4GB, I focus on only one of the peripheral subgroups in the society. Its total population is small relative to the dominant group. There is only one basic genetic strain, although there is some mild ethnic variation within this group. The concentration of authority within this group ranges from medium to high. This is a simple homogeneous culture.

The ideology of this group is belief-oriented. This ideology is pervasive within the group, and its strength ranges from medium to high. This group is in a state of constant struggle. Some group fragmentation occurs within this sub-macro unit. The institutionalization of behavior for this group is both high and enforced in some fashion. While the group's activity is predominantly focused on the physical realm, its use of telepathy is somewhat greater than that of the dominant group.

A significant psychological difference between this group and the dominant group is that the subspace side of this sub-macro group has a sense of purpose and, indeed, a mission. But the physical mentality of this group does not recognize its subspace counterpart any more than the dominant group.

There is tension between this group and the larger society. This group has a place within the larger society, but it must defend its position constantly. On the other hand, the macro-society recognizes the place of this sub-macro group, and it grudgingly accepts the fact of its existence and strength.

The leader of this subgroup has an intense emotionality. He has a strong relationship with his group. Group members generally recognize his privileged position of authority, and they believe that this authority is conveyed from a higher source, as with a belief system.

Shifting to a typical non-leader member of this sub-macro group, this person has a strong sense of group identification. This person's concern for the success of his group is very high, and he has almost tunnel vision with regard to his support of his group's needs. This person has a weak recognition of the macro-society's needs. Success of the group is more important than that of the macro-society. The dependence of the group on the larger society is only marginally recognized.

Executing a consciousness map for this sub-macro group, I perceive on the physical level that the defense of the group's concerns and agenda is paramount. Interestingly, the subspace psychology is also very supportive of this physical-level agenda.

To summarize the Phase 4GB data for this sub-macro group, I perceive that this group is securely located within the macro-society. It defends its interest both on the membership and leadership levels. It is committed to support the macro-society, but only as a function of its own ability to thrive within that larger arena. Collectively, the group recognizes some indebtedness to the macro-society, but this is not deeply felt. The group's identity and its highly patterned behavior is predominantly historically based.

The Target Society's Developmental Trajectory

Moving to Phase 5, I explore the macro-society's developmental trajectory. In the beginning of that trajectory, there is a significant starting point that is culturally defined. At some point in this society's history arose new ideas that precipitated a different sense of culture and communal identification. Shifting to the end of the developmental trajectory, I perceive what could best be described as a "termination point" for the definition of the macro-society. The social structure is held together by institutionalization at this point, but not by the force of the original ideas that formed the genesis of this culture. Between the beginning and the end of this developmental trajectory, I perceive general continuity. However the initial precipitating ideas that were relevant in the beginning of this trajectory wear thin near the end, and the decay seems gradual.

Prior to the beginning of this trajectory, there appears to be considerable turmoil, struggle, and conflict, most likely violent. There is the breakdown of a previous social structure. At the end of this trajectory there is also conflict, but it is not violent, or at least not very violent. There is a challenge to the historically dominant authority structures that results in the overthrow of these structures.

Blind Analysis

Following the completion of my session, I was given the following list of essential cues for five targets. As is typical of blind analysis, I needed to order this list, giving highest priority to those targets that most closely fit the session data.

North American Convention of Baptists (current time)
Roanoke Colony (one year after establishment)
Oxford University Faculty and Students (circa 1900)
North Korean Society (current time)
The Holiday Inn Staff (current time)

During my analysis of the data, I found very little support for the Holiday Inn target. I was familiar with the hotel since The Farsight Institute maintained its offices there for a year. While there are some divisions within the staff in the hotel, these divisions are predominantly language- and ethnic-based, which is contrary to the nature of the data that I obtained in my session. I also rejected the Oxford University target because the sense of social control by a dominant leader and group that I perceived during the session does not correspond well to the academic atmosphere of a university. The Roanoke and Baptist targets were better supported by my data, since they both contained a single ethnic type. But the Roanoke Colony was too small, and my data suggested a connection between the physical and subspace realms that was too poor to support a convention of Baptists, who spent much of their time praying. The North Korean Society target seemed to fit the data perfectly, however.

In North Korea there is a single dominant political leader and party. The leader has just recently solidified his control over the political infrastructure that he inherited from his deceased father. Political insiders have noted that this leader spent a considerable period of time consolidating his hold on power, and that he seems adept at manipulating the power struggles within his society to his benefit. North Korea is also experiencing extreme environmental and economic conditions at the current time. There is a high degree of malnutrition, and even starvation, on a mass scale. The central authorities within the country are not being challenged severely. There is only one ethnic group, and the commonality that connects all North Koreans is cultural. A close examination of the remainder of my data suggested a very close match with the North Korean target, and I was informed that the target for the session was in fact North Korean Society (current time).

Discussion

The complete essential cue for this target is "The social and political systems of North Korea (current time)." This example demonstrates that social scientific data can be obtained using remote viewing. These data also give us some insight into a possible future for North Korean society. My interpretation of the Phase 5 data (the society's developmental trajectory) would lead us to expect a change in the current government at some point in the future. This change may indeed be radical. It seems as if the society is going to enter a period of direct challenges to the dominant authority structures. This may mean the end of the current communist system in North Korea. On the other hand, it could also be that the current set of leaders as well as the system of government will be forced to change significantly, but not necessarily to a revolutionary degree.

Remote viewing is not limited to physical descriptions of people and places. Our task as students of this new science is to explore and invent innovative ways of communicating information between our physical and subspace minds. These new ways will allow us to extend our current capabilities to obtain useful information. But they will also structure our own education as we increasingly understand this universe of mystery.

Readers who want to see additional SRV sessions of verifiable targets are encouraged to visit the website of The Farsight Institute at www.farsight.org. The website has actual handwritten SRV sessions freely available to the public, and even recordings of monitored SRV sessions as they took place.

SECTION TWO

Chapter 13

What We Know and Don't Know about Extraterrestrials

We've learned that remote viewing is not limited to perceptions of only human phenomena. What blinds us from continuous awareness of everything other than our own existence is the genetics of our human bodies that limit our perceptual abilities. Remote viewing helps alleviate this limitation.

In a situation without perceptual limitations, there is absolutely nothing stopping a remote viewer from targeting nonhumans who do not live on this planet. Once you become accustomed to remote viewing, you also become accustomed to the unusual nature of the information that can be obtained. It it not wise to disregard some information as esoteric or unverifiable just because one does not yet have physical corroborating data. Indeed, remote viewing using verifiable targets is practiced in order to develop the skills necessary to obtain information from targets about which little or nothing is known. Once such remote viewing data is obtained, the problem then shifts to doing whatever is necessary in order to obtain complementing physical evidence.

In a previous book, *Cosmic Voyage: A Scientific Discovery of Extraterrestrials Visiting Earth,* I presented remote-viewing data that strongly support the idea that a number of extraterrestrial cultures are interacting with Earth at the current time. This chapter contains a brief summary of these basic findings. Any possible errors in the interpretation of my remote viewing data are my own, of course. But others have seen what I have seen.

At The Farsight Institute, we have had many viewers target extraterrestrials under a variety of solo, single-blind, and double-blind conditions (that is, data types 3, 4, and 5). The quality and quantity of corroborating data are such that I have decided that the scenarios involving Martians, Greys, and others are most likely real.

I am fully aware that the proposition that extraterrestrials exist on Earth appears ludicrous to many people. But we have an interesting dilemma. These remote-viewing sessions must be explained. We cannot simply dismiss countless replications. We either have to explain a psychic mechanism that could produce such results repeatedly, or entertain the proposition that we are living in a collective state of denial regarding life that appears to have evolved millions of years ago even in our own solar system. So examine the remote-viewing data carefully. Then readers should ask themselves what they think is the truth.

According to some of the basic findings, there was once an ancient Martian civilization that was destroyed by a natural disaster involving a comet or asteroid. A highly advanced group of extraterrestrials whom we now call the Greys arrived on that planet with the approval of a loosely organized galactic organization (the "Galactic Federation") soon after the disaster and rescued the surviving Martians. Due to the Greys' technological capabilities (their ships travel across time as well as space), they were able to bring many Martian survivors to our current time period. The Martians were established in temporary underground colonies on Mars and given a limited level of technology that allowed them to sustain their now relatively small population. They also have the capability to fly brief interplanetary trips to Earth.

Mars is essentially a dead world except for these refugees. The goal of these Martians is to transfer all of their surviving population to Earth. Through remote viewing, one underground facility was located in New Mexico, apparently under a mountain named Santa Fe Baldy. This facility seems to be used as a center for operations while Martian-controlled ships deliver supplies to surface villages on Earth containing groups of Martian emigrants.

The Greys also have an interesting history. They come from a home world not in our solar system. They destroyed their own world due to rampant and selfish abuse of their environment, and were forced to move their civilization underground. They then began a long process of manipulating their genetic code to allow them to adapt to their new conditions. They developed large eyes that require very little light to see. They allowed their bodies to shrink to the point that they no longer could experience live birth, requiring their offspring to be "born" using artificial containers (that is, fetuses grown in canisters). Finally, the Greys manipulated the genetic code for their physical minds. Fearful of their past, they rid themselves of volatile emotions. They could no longer feel anger, hate, or fear. Unfortunately, they also lost the ability to feel positive surface-level emotions such as love and compassion.

But the journey of the Greys into emotional neutrality was not without benefits. This species made a collective decision to advance spiritually. They explored both physical space and subspace, searching for meaning to their existence. Ultimately, they turned their attention toward God, not a fuzzy God that is the object of pleading prayers, but a real God with a consciousness and, indeed, a personality. They sought to understand the physics of God as well. To my understanding, there was no question about God or spirituality that they were afraid to ask.

During their evolutionary journey toward God, they met some powerful personalities that impressed them greatly. They realized that they needed to experience certain things in order for their own evolution to mature. They were caught in a dead-end road, and they needed new physical resources in order to advance further.

Returning to their former genes was never an option for the Greys. They held a deep fear of their past. If their former genetics led them to an existence in which they selfishly destroyed their own planet, what guarantee could be made that they would not do so again? The Greys wanted a different route to the future, a route that required outside help.

Apparently after serving the Galactic Federation as members in good standing for a long time, the Greys applied for permission to travel to Earth and obtain human genetic material to create a new race for their souls to inhabit. My data suggest that they have carefully followed a policy of asking permission of human souls before physical birth with regard to their participation in the Greys' genetic activities. With permission (on the soul level) granted, the Greys have been visiting many humans (often at night) for the purposes of working with human biology to create a new race of Greys, a race that would again be able to experience live birth and to feel emotions (primarily positive ones).

This is where *Cosmic Voyage* left off. In retrospect, my earlier book turned out to be an essentially accurate but naive journey into the realm of cosmic explorations using these new tools of consciousness. After finishing *Cosmic Voyage*, I was left with the impression that the universe was filled with good ETs, and humans needed to be chastised for not recognizing the world of wonder that waited so patiently above us. Only after many months did I realize how incomplete this story was. Now as I look up into the skies from my backyard, I see not a heavenly paradise, but a galaxy of conflict, growth, and mystery. What I have come to learn is that there are indeed good ETs who want the best for humanity. But there is a struggle going on at the current time, and there are enemies to a human destiny that fights to Be.

There is a war in space and beyond. This war has just now reached our borders. What is at stake is not the awakening of a foolish humanity that wants to keep its head buried in the sand, but a future for our children that is free from fear, imprisonment, and possibly genetic slavery or even genocide. We are entering a new stage in a long battle in which we must fight for the survival of our species. We are now living in an age that requires courage above all things. If fortune favors the bold, so must be our destiny.

PART III

THE WAR

Chapter 14

The Mars96 Russian Space Probe

While conducting research at The Farsight Institute in 1996 and 1997, I came across some information that suggested that the space probe launched by Russia named Mars96, may have been deliberately destroyed. Officially, the probe malfunctioned soon after launch and was destroyed when it fell back into the Earth's atmosphere. We tried some experiments at the Institute with many of our advanced remote viewers in order to determine the cause of the probe's destruction. Oddly, we found that none of them could lock on to the target. Indeed, some of our viewers reported feeling blocked by some outside force during their sessions. This experience occurred repeatedly, and nothing we could do seemed to allow our viewers to perceive this target.

Around the same time we also attempted to target a group of reptilian ETs that we had been hearing about from a variety of different sources. As with the Mars96 targets, the viewers had unusual experiences, and many of them felt as if they were being blocked in some fashion from perceiving the target. Indeed, some viewers stated that someone or some group was aware of them, and that an active attempt to interfere with the remote viewing had occurred. Interestingly, many viewers reported headaches both during and immediately after viewing these targets. Some viewers wrote on their sessions that they felt they were being attacked.

Nothing like this had ever happened before. We had always been able to target whatever we wanted. We marveled at the idea that anyone could interfere with what we perceived to be our God given right to poke our nose into anything we desired.

Soon afterward, the suggestion was made that we develop a technique that would use outside help to eliminate the interference. The only beings that I thought might be willing to help were those in the Galactic Federation on whom I had previously reported in *Cosmic Voyage*. We were probably the only group of humans trying to communicate with these ETs, and they might be willing to assign someone to be our official contact person if we asked them. We could attempt going through the consciousness of this being to the target, leaving it up to the ET to clear the path. Having no other alternatives, we decided to give the idea a try.

The remote-viewing sessions began by initially targeting the Galactic Federation contact person. The viewer then would enter the mind of this person and exit at the correct target location. More advanced forms of this procedure have been used for some of the sessions reported in this book, in which the navigation through the consciousness of the Galactic Federation contact person is accomplished in the target cue, to which the conscious mind of the viewer is blind.

To my great surprise, nearly all of the viewers performed very well with the new targeting method, regardless of the type of target attempted. With regard to the Mars96 probe, the viewers perceived a space probe that was disabled and that fell back into the atmosphere of a planet. However, some viewers also perceived that the event that disabled the space probe was not an accident. These particular sessions did not have a great deal of detail regarding this event, but it seemed possible that some outside ET group was involved. Many viewers did perceive that this outside ET group was associated with the color orange.

The reptilian target was equally interesting. In this case, the viewers clearly perceived a group of reptilians. The "Reptilians" were acutely aware of our remote-viewing efforts, and they were very disturbed that we somehow managed to perceive them. Many of our viewers had mild headaches during the sessions, and nearly all of them perceived that the Reptilians were extremely surprised when we "popped in."

In the remote-viewing session presented in this chapter, I was given a target relating to the Mars96 Russian space probe. The advanced form of cuing is used in which I was blind to the use of the anti-blocking procedures. The essential cue and qualifiers are as follows:

Target 2935/7923

Protocols used for this target: Enhanced SRV

The viewer perceives through the consciousness of the Galactic Federation Contact Person for The Farsight Institute, to remote view the destruction of the Mars96 Russian space probe. In addition to the relevant aspects of the general target as defined by the essential cue, the viewer perceives and describes the following target aspects:

- the physical cause of the destruction of the Mars96 Russian space probe
- the object that is left in space following the destruction of the Mars96 Russian space probe, as well as the purpose of this object, if the placement of this object in space is causally connected to the destruction of the Mars96 Russian space probe, and if the object exists
- the technology that is used to destroy the Mars96 Russian space probe, if it is purposefully destroyed by technology

Note that this target cue is somewhat open-ended. Some cues have to be open-ended when the basic descriptive parameters of the target are not yet known. The cue is designed to let the subspace mind determine the most important aspect of the physical cause of the destruction of the Mars96 probe. For example, the cue does not state that the viewer must perceive the piece of metal that may have broken apart when the probe first experienced difficulty. The cue also does not state that the viewer will perceive the heat that incinerated the probe as it re-entered the atmosphere. The cue leaves it up to the subspace mind to locate and to determine the information that we need to know in order to explain the physical destruction of the probe. It does, however, permit a focus on any technology that may have been associated with the destruction of the probe, as well as any possibility that some object may have been left in space by the probe that may be causally connected to the probe's destruction.

28 April 1998
1:20 p.m.
Atlanta, Georgia
Protocols: Enhanced SRV, Type 2
Target coordinates: 2935/7923

I initially focus on a medium-build male nonhuman being that has an orange skin tone, and I deduct a Reptilian. I also perceive a different being who is short and has a pale or gray skin tone. This second being is wearing a body suit, and I deduct that he is a member of the Greys. I also perceive a subspace being who is surrounded by a circular shape that is emitting light. Finally, I perceive a structure that contains multiple levels, as well as many beings. I draw a large picture of the exterior of the structure in my Phase 3 sketch.

In Phase 4, I perceive that there is an enormous subspace component to this target. Indeed, the physical and subspace sides of this target feel about equal in their representation. I perceive many subjects that resemble Grey ETs. I also perceive light coming from a large circular object or globe in subspace, and I deduct the Galactic Federation Headquarters.

I move slowly toward the globe object, and I find myself sinking deeper and deeper into some type of dimensional shift. The global object is acting much like a center of gravity, subspace gravity, like a black hole, but white. It feels like a place that I have been before.

I move to the location in or on the globe object from which I am to observe what is important to the target. I now feel like I am in the Galactic Federation Headquarters structure. I am in a large chamber. There is a light everywhere, and there are subjects in this chamber. I move to the optimal location inside this chamber for satisfying the informational needs of the target cue. There is a subject here, and he is seated and facing me. I perceive that I am supposed to move into the mind of this subject, and I do so.

I emerge at a new target location. There is land and vegetation here. This place feels more physical than the globe object, but not as much as the Earth. On the subspace side of this location, I perceive chaos and turbulence. There is a war going on. There is a battle, with shooting, guns, the works, and ferocious fighting. This is a no-survivors type of war. There is confusion, anger, and fear. The emotions of the entire place are mixed up, upset. This feels predominantly like a subspace battle. The physical side of this location seems calmer than the subspace side. Yet some fighting clearly is spilling over the physical/subspace divide.

What is happening now is the culmination of something that began long ago. This is an event of huge magnitude. The emotions seem to reflect two sides to this conflict. There is a defending side and an aggressor side. While the heavier physical realm seems to be away from the center of the conflict, subjects on that side are nonetheless watching the struggle that is going on with great interest. It is hard to underestimate the enormity of the subspace battle and struggle.

The aggressor side has many spaceships. This side is planning its activities thoroughly. There is premeditated coordination and ruthless execution. I execute a collective deep mind probe on the aggressor side. They are driven by ideas, but are very afraid of something. They are afraid of something in themselves, and they are striking out to stop it by destroying something that will force some type of acceptance of who or what they are. Fear is the dominant emotion driving the aggressor side.

Cuing on the purpose of the activity by the aggressor side, I perceive that there is no acceptable alternative other than to fight, to conquer, to destroy, and to force conformity. It does not feel like these subjects have thought through their actions very well.

Shifting my focus to the defending side, I perceive that they are more passive. They do not want this conflict. The defending side feels that they are morally or ethically correct. They are resolute in their determination to fight and to defend themselves.

I then cue on what is motivating the aggressor side to attack. The defending side has something, or they have taken something. The defending side has something that the aggressor side wants. It is a prize, like a "jewel" in the minds of many. I deduct the Earth.

It is an entire civilization, a planet. The fight is over the control of an entire world. It is as if the defending side has obtained some type of agreement, possibly fairly, and the aggressor side lost out and is now attacking to destroy.

I move to the planet that is the focus of the conflict. The planet feels more physical than subspace. It has blue water, and from space it looks beautiful. The place looks and feels a lot like Earth. I am viewing this place now as a beautiful and prized jewel of an ecosystem. It is a world with a rich and lush ecosystem. This place has tremendous biodiversity. It is rich in terms of its environment and its resources, both biological and physical.

I move to the aspect of this world that is most valued by the warring parties. I perceive subjects that feel very human. They are a resource, almost a commodity. They are a gene pool.

What is at stake is the ability to exploit a gene pool that has extreme value to the warring parties. To the warring parties, access to this gene pool is seen as an extremely important advantage. It is as if the aggressor side would feel deprived of something it needs, something that would give it a strategic disadvantage not to have. The aggressor side feels that fighting is the only way to ensure what it perceives to be its own right to evolve.

The aggressor side does not feel that it would be guaranteed access to this gene pool if it went along as a cooperative player with other groups. These subjects have somewhat of a "self-esteem" problem. They feel like they are the "odd man out." I perceive that they have had some problems cooperating with others in the past.

Discussion

This is why we sometimes leave cues open-ended. From this session it seems clear that there is a war going on, which in turn seems to be the ultimate cause of the physical destruction of the Mars96 Russian space probe. There are many questions yet unanswered with regard to this target. It appears possible that some of the war-related technology was used to destroy the Mars96 probe. Yet we do not yet know the group responsible for this event. From the initial data, the conflict seems to be between the Greys and the Reptilian ETs. Yet it is not yet clear who are the aggressors and who are the defenders.

From this session, we do not know why the Mars96 space probe got caught up in this struggle between two warring groups. Apparently the battle is occurring nearly entirely in sub-space, and humans are totally ignorant of the conflict. I can only assume that the Mars96 probe would have assisted one side of this conflict in some fashion, and that the other side needed to destroy it. It seems as if both sides are taking pains to keep the conflict hidden from Earth humans. Moreover, both sides seem to be working in a clandestine fashion along a variety of dimensions relating to Earth. This session raises more questions than it answers, but it does answer one important question. My interpretation of these data suggests that the destruction of the Mars96 Russian space probe was no accident. It was a casualty of war, a war that may eventually have great consequences to life on Earth.

Chapter 15

Those Who Destroyed the Mars96 Russian Space Probe

The next session answers the question raised in the last chapter of who destroyed the Mars96 Russian space probe. The essential cue and qualifiers are as follows:

Target 2947/8924

Protocols used for this target: Enhanced SRV

The viewer perceives through the consciousness of the Galactic Federation Contact Person for The Farsight Institute, to remote view the subjects responsible for the destruction of the Mars96 Russian space probe (at the time of the physical destruction of the Mars96 Russian space probe). In addition to the relevant aspects of the general target as defined by the essential cue, the viewer perceives and describes the following target aspects:

- the ET group directly or indirectly responsible for the destruction of the space probe, if such a group is responsible for this destruction
- any humans or human organizations that passively or actively participate in the destruction of the Mars96 Russian space probe, if such humans or human organizations exist
- the primary thoughts of subjects responsible for the destruction of the Mars96 Russian space probe, if such subjects exist

Note that the target cue is again somewhat open-ended. If those responsible for the destruction of the probe were drunken technicians who did not launch the probe correctly, then such information would be acceptable to the cue. However, if the probe was destroyed intentionally by humans or ETs, the cue would allow for the perception of this information as well. Especially note that the target cue does not say anything about where to view the subjects. If the subjects are ETs involved in a subspace battle, as suggested in the previous chapter, then the session will go there. The cue was written to allow the subspace mind to determine which information would be appropriate.

7 May 1998
2:40 p.m.
Atlanta, Georgia
Protocols: Enhanced SRV, Type 2
Target coordinates: 2947/8924

I am perceiving a spongelike artificial structure. It reminds me of the catacombs of ancient Rome, although I am not certain whether it is a physical or subspace structure. I also perceive a nonhuman subject who has orange and green tones to his skin. I deduct a Reptilian. This subject's skin has a patterned texture.

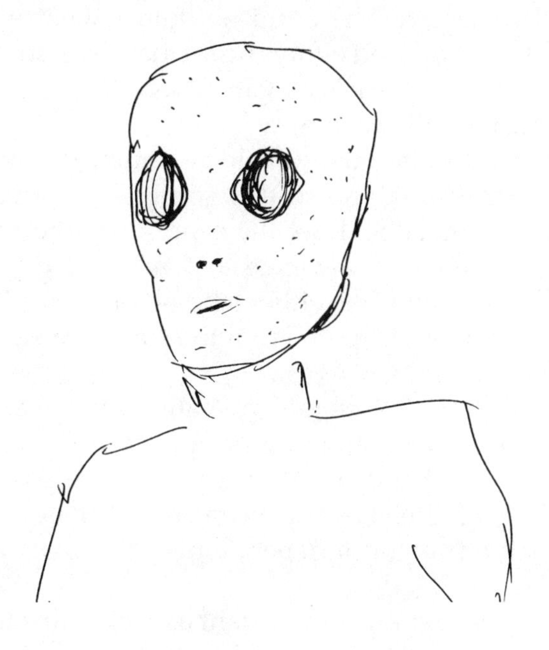

There are also high levels of energetics at the target site. I detect splashing, waves, and something wet. I also perceive strong upward energetics, and I sketch something that looks like a mushroom cloud. I deduct an atom bomb.

Moving on to Phase 2, I hear voices and winds, and I perceive something splashing. The tastes are salty and the smells are of fish. Something is tall and towering as well as wide and curving at the site. I perceive fast and expansive, even explosive, energetics. In Phase 3, I draw a large sketch of what appears to be an explosive mushroom cloud.

In Phase 4, I again perceive something circular and rising with explosive energetics. There is bright yellow light. I perceive the sense from subspace of the need to evacuate an area. It is like subspace life has been cleared out of the area, and a vacuum of life remains. Something in subspace is scattered. The event is as destructive as it is surprising. It is a large-scale event. It leaves subjects numb and in shock. I sense the concept of eradication. It is like the people are gone. They have been removed, killed, or wiped out.

The feeling is one of having pests removed. There is tremendous yellow, hot energy, and I clearly perceive a sense of innocence among the subjects who have suffered here. It is like the subjects who once lived here did not know what was coming. There was deception and ignorance. I perceive structures and streets, and I deduct New York City. (Note: This is a deduction, not data.) This feels like an empty city.

I move to the location that would be optimal for me to understand what happens to cause this scene. I perceive an object that is hard and metallic. I get the concept of a terrorist whose emotions are filled with revenge and anger. I draw a sketch of something that looks like either a bomb or rocket. There is the concept of a terrorist weapon. From the subspace side I perceive wariness.

There is a physical subject, a male. He is wearing khaki clothes, and he has a paramilitary appearance. He is foreign, in the sense of not being an American. This person has a poor understanding of life. He has been indoctrinated and is not thinking analytically or independently. His plan is to cause disruption.

I move to the next most important aspect of this target, and I perceive air and sky. There is a patterned set of objects moving horizontally near the horizon. I am currently positioned in the air about a mile up, near some white clouds. Moving close to the objects, I perceive that they are flying quickly, and I deduct UFOs. I move into the most significant moving object and find it to be metallic and hollow. There are subjects seated inside the object. I get the sense that I am not supposed to focus on the mentality of the subjects, but rather to observe their activity. They are moving things—like levers—with their hands. Everything is happening very quickly inside this object.

I move to the location that would be optimal for me to understand why I am perceiving this scene for the current target. I am now in a confined area with one subject. He seems to have a beard, and he is holding some device. I am in a hollow metallic structure that feels like it is made of a light material. The subject is alone, and I sense that he is flying, and perhaps he is hiding, or even hiding something other than himself.

I move to the center of the target and again perceive a large explosion with a mushroom cloud. I quickly shift to the next most important aspect of the target and find a male subject. There are tremendous levels of orchestrated activity in subspace. Someone is "calling in the reserves." There has been a calamity, and I feel the general sense is one of shame. This did not need to happen. The male physical subject is a terrorist.

I move 2,000 feet above the location of the explosion at a time 10 minutes before the explosion. There is a congested physical city. The time seems to be in the afternoon. There is smoke and smog, as well as city noises of all types. The emotions of the people in this city are life- and work-related, a normal emotional mix.

I move to the primary event for this target. I am now very high in the atmosphere over the surface of a planet. I can perceive the horizon. It is dark. This does not look like Earth. The colors are too purple and black. In the subspace realm, subjects are watching a fight. There is a war, a tremendous struggle in progress.

I move to the center of the struggle. There appear to be Greys here. These are early Greys. They are involved in some kind of fight that uses advanced technology. I cue on the purpose of the struggle. It is not well thought out. Much of what is happening is reactive and due to a lack of communication.

Moving to the location that would best help me understand the connection or relation between the various major themes of this session, I find a subject who is a terrorist. I perceive that the terrorism is reactive and not well thought out. This is a common experience among many species, perhaps most—or even all—species. There is a need to move past this juggernaut in social evolution. For things to improve, it is necessary to move beyond this point of tension in linear evolution.

Discussion

This session has to be considered together with that in the previous chapter. Both sessions contain data suggesting that a state of war exists between two large groups. This war appears to exist in a location that is dimensionally different from our own. It is a subspace confrontation.

I perceived the Greys in both sessions, but they did not appear to be the aggressor side of the conflict. The Reptilian ETs were also perceived in both sessions, and the flavor of the data seem to suggest that the Reptilians are the aggressors in this war. The Greys seem to be defensive, even moralistic, in both sessions.

The Reptilians may have collaborated with a few humans in an overall aggressive strategy that included the destruction of the Mars96 Russian space probe. I can only speculate as to why the Reptilians would be interested in destroying the probe. The answer may be in the nature of Mars itself. Based on remote-viewing data presented in *Cosmic Voyage*, recall that the physical ruins of the ancient Martian civilization still exist on the surface of that planet. Moreover, survivors of that ancient civilization still struggle to survive on their nearly dead planet and in hidden retreats here on Earth. Perhaps the Reptilians have an interest in keeping humanity in the dark with regard to extraterrestrial life. They may have an agenda that would be seriously compromised should information about their activities, and the activities of other extraterrestrials, become well known. This is speculation. It will be further examined in the sessions yet to come.

Chapter 16

The Reptilian Extraterrestrials

Who are these Reptilian extraterrestrials who are fighting this subspace war? Before beginning this project, I had hardly any information about the Reptilians. Given no leads that would suggest concrete targets on or near Earth, the best way to proceed is with an open cue that simply targets the Reptilian ETs. In this way we leave it up to a viewer's subspace mind to locate the information that will be most helpful in the current context.

The essential cue and qualifiers are as follows:

Target 3095/2934

Protocols used for this target: Enhanced SRV

The viewer perceives through the consciousness of the Galactic Federation Contact Person for The Farsight Institute, to remote view the Reptilian ET species that is currently operating on or near Earth (at the time of tasking). In addition to the relevant aspects of the general target as defined by the essential cue, the viewer perceives and describes the following target aspects:

- the physical environment of the subjects' living conditions
- the age and gender variations among the subjects
- the emotional state of the subjects
- the dominant groups among the subjects, including any governmental organizations
- the primary thoughts of the collective consciousness of the subjects
- the level of technology available to the subjects

Note that the target cue is open-ended. It does not say anything about the location of the subjects. It says only to view them at the current time. If this species truly is involved in a subspace battle, then that may be what I will perceive. On the other hand, if I am wrong about the battle, or if my understanding of the situation is incomplete and I need to view something else to change or to enhance the picture, then the subspace mind will locate my perceptions elsewhere.

8 May 1998
3:04 p.m.
Atlanta, Georgia
Protocols: Enhanced SRV, Type 2
Target coordinates: 3095/2934

In my initial approach to the target I perceive a subspace being surrounded by yellow and white light. I perceive that this being wishes me to enter his consciousness, and I do so. I immediately emerge at a new location. There are multiple artificial structures here. The colors are dark green and orange. The smells are sooty and smoky, and I sketch what appears to be a city landscape. I focus on a single structure and find that it is crumbling. I again perceive the smells of smoke and the sense of soot. I also perceive the concept of fighting. With my next ideogram I perceive a subject. The subject is in pain. He is burning and

writhing. I also sense that the location is foreign. My next ideogram suggests dark, dirty, and sooty air that is filled with smoke and putrid smells. My Phase 3 sketch is of a mushroom cloud exploding over a burning city.

In Phase 4, I again perceive a black, sooty, and dirty environment. Everywhere there is pain, flames, writhing subjects, and burning flesh. I perceive the flesh burning right off a subject's bones. The fire is incinerating them. They are crying out as their flesh continues to char. Everything is black here. Cinders are everywhere. There are sounds of crying everywhere, some loud and some soft. There are babies crying too.

I sense heroism here, amidst this defeat, but it is futile. Heroism cannot change the outcome. Only death lives here. This is a city, or what used to be a city. It is now a holocaust. This is what I would expect a city to look like after an atom bomb blast.

There are legions of beings in subspace. Subspace workers are collecting the souls of the dead. They are trying to help, but basically they are just getting these souls out of the situation. This is an evacuation. Everything is in a state of collapse. The subspace workers are letting it go, abandoning this place forever. Whatever was here has met its final end. This feels like a surrender. Whoever these beings are, they are giving up and getting out.

My senses are again assaulted by the reality of this holocaust. The charring, the pain, the smoke, are all I see.

I move to the next most important aspect of this target. I am in a purely subspace location now. There is the sense of regrouping here. There is a meeting, and a small group of subspace beings are gathering. I sense the number of subjects to be around 12. When I probe the physical column of the Phase 4 matrix, there is the backdrop of the charred environment. But when I probe the subspace column, I return immediately to the meeting. I briefly experiment with probing the physical and subspace columns as I become familiar with this binary split, and as I do so, I become aware that the subspace side is the important part of the session.

I perceive a central male subject. This is a council of some form, and this is one of their formal meetings. The group reminds me of a governing body, like a town council, and I have the guided deduction of the Galactic Federation. I perceive that I should enter the mind of the central subject, and I do so.

I emerge at a new location. I am in space, and a planet is in front of me. It looks like Earth. It has blue oceans and apparently clean air. I sense that this is Earth, an Earth that beings in subspace see as a gem. I also sense that I am receiving information from the central subspace subject while I watch the Earth. I move around the surface of the planet. There is life here, buildings, bustling people.

I am supposed to perceive this location from the perspective of a subspace being. This place is alive, a place to visit that is filled with adventure, varied experiences, and a wide range of personalities. It is like a kitchen where things are made or prepared. It is a place of deep value in some way that I do not fully understand.

I move to the next most important aspect of this target, and I again perceive the group of subspace beings in their meeting. I execute a collective deep mind probe on all the beings in the meeting. From their thoughts I perceive that there needs to be an effort on the human side. Subspace cannot force the events on the physical side to go one way or the other. There must be action on the physical side. Nothing can be done except to teach. Teaching by example is the way to influence the physical.

I sense one other important thing that I need to perceive in this session. I move to the optimal location. There is a male physical subject. I am being drawn into his mind. From the perspective of his consciousness I am supposed to describe the nearby environment.

There is a structure on land near water. This place feels like a blend between physical and subspace. It is like both realms are together, transparent to each other. Everything, the life, the consciousness, even the land is one. There is harmony here.

I cue on the question of where this place is. It is in the future, our future. This is a place where people are trying to go. It is a goal, though temporary. This future point is attracting the attention of the current time frame. It might best be described as a lighthouse that is shining a beacon to those who struggle in stormy seas.

Discussion

As with many remote-viewing sessions, the range of these data appears to address realms beyond the constraints of the essential cue. The first part of the session clearly corroborates and enhances the previously presented data that suggest that the Reptilians are indeed involved in a major military confrontation. They seem to have suffered many casualties.

It is not easy for me to interpret some of the data. If the subspace council meeting is associated with the Galactic Federation (which is my suspicion), then the second part of the session acts as an interpretive guide to the first part. That is, the scenes from the nuclear-like holocaust depict some of the current experiences of the Reptilian species. The second part of the session tells us either that the Reptilians view the Earth as their own Shangri-la, a prize worth fighting for, or, alternatively, that humanity on Earth at some future time is trying to influence our current behavior to evolve in their own more harmonious direction. If the second scenario is correct, then we can either act in a way that leads to the nightmare of the Reptilian conflicts, or to a future in which we live in peace and harmony with the greater subspace realm.

I lean toward the latter interpretation. I suspect that the Galactic Federation felt that they were permitted the discretion to offer a stark contrast. I now have no doubt that the Reptilians are fighting a war in which they are either causing or experiencing heavy casualties. They also seem to be the aggressor side in this conflict.

Yet I still do not understand why these beings find Earth so attractive. What do they want from this planet? What do they want from us? How do we humans fit into this conflict, and how can we head in a direction of safer shores?

Some of these questions are answered in the next session.

Chapter 17

Reptilian Activities on Earth

If there are Reptilian ETs on and near Earth, what are they doing here? What are their activities, their operations, their agenda? Are they interested in the animal life, the human life, the physical resources of the planet, all of the above, part of the above, or what? Since we have few specifics on which to base a cue, the open-ended approach is optimal in this situation. The session for this chapter probes the activities of the Reptilians, as well as some psychological matters.

The essential cue and qualifiers are as follows:

Target 7394/3908

Protocols used for this target: Enhanced SRV

The viewer perceives through the consciousness of the Galactic Federation Contact Person for The Farsight Institute, to remote view current activities on or near Earth of the Reptilian ETs in which humans are involved (at the time of tasking). In addition to the relevant aspects of the general target as defined by the essential cue, the viewer perceives and describes the following target aspects:

- a location that optimally reveals the most information regarding the Reptilian ET activities involving humans
- the psychological and emotional state of the humans interacting with the Reptilian ETs

- the psychological and emotional state of the Reptilian ETs during their interactions with humans
- the purpose of the Reptilian ETs' activities involving humans

6 May 1998
11:28 a.m.
Atlanta, Georgia
Protocols: Enhanced SRV, Type 2
Target coordinates: 7394/3908

I decode my first ideogram as a hard man-made hollow structure with an irregular oval topology. My second and third ideograms represent a mountain. The fourth ideogram is of a subject. There are orange and green colors, and I deduct a Reptilian. My final ideogram again represents a humanoid subject. The colors are beige, green, and orange, and I deduct a young Reptilian/human hybrid being. In Phase 2, I hear voices, and in Phase 3 I sketch what appears to be a mountain with an underground structure within it.

In Phase 4, I focus on the interior of the structure. I deduct a facility, and it has the flavor of a cavity inside of something. I am inside the structure and it is very dark. The room that I am in is either square or rectangular in shape. There are squarish objects on the floor the size of moving boxes. There are many other things and objects in the room, and I deduct a storage room. In the subspace realm I perceive the emotion of disapproval and being wary. This feels like a central place of operations, and I have the guided deduction of a Reptilian base.

The emotionals of this location do not feel human, nor do they feel like Grey emotions. They are different, and they feel somewhat new for me. I shift to a time in which there is activity in this room, and I perceive dim light and subjects who are working. The subjects feel like scientists. They are moving carefully and slowly while they are doing something. The skin of these subjects has a surface to it that is harder in some spots than others. There is a pattern or texture on the surface as well. They are wearing clothes that are unusual in some way; the clothes may be one-piece or skin tight.

I again note that the subspace realm conveys a sense of being wary about this place. I cue on the purpose of the activity at this location and I perceive it to be task-oriented around a specific assignment. These subjects are not spending idle time.

I move to the next most central aspect of the target, and I perceive light and fast energetics. There are humanoid subjects watching experiments of some type. The subjects have more human emotions. Though they seem human, something is not right. It is as though they are half-human. The lower part of their faces seems flesh-toned and more humanlike, but the upper part of their faces are different, patterned, textured. These beings seem like young adults, and I deduct hybrids.

It feels like these new subjects are graduating from some kind of training, like college, and they are beginning new courses. They feel special and comfortable in their surrounding. I deduct "a chosen race." There is not much light in the room, and I get the sense that these subjects need and enjoy more light. But the light is somewhat dim because of the presence of others.

I move to the next most important aspect of the target, and I find myself outside looking at land, sky, and rocks that lie on rough and steep slopes. The air is clear and crisp. It is windy, but the weather is nice. The sun is out. I move 1,000 feet up from my current location and I find myself looking down at mountains and trees. These are low mountains. They remind me of Santa Fe Baldy, but they feel different somehow, like farther west.

I then execute a movement exercise where I slowly shift from my current location to the interior of the first structure that I described earlier in the session (sliding—see Appendix 2). I find myself passing through air, then rock, seemingly traveling for a while. I am now in that intended location, and I note that I moved downward and diagonally, followed by horizontal movement to arrive at my new location.

Discussion

From a remote-viewing perspective, this session went very well. The cue was designed to have me "drop in" to an appropriate location and simply observe what is going on at the current time. The cue was not designed to seek out any particular type of activity.

It seems as if mountains are ideal locations to hide underground bases. In *Cosmic Voyage*, I report about a group of Martians who have a base in the United States underneath a mountain in New Mexico (Santa Fe Baldy). Secret human military bases are found under mountains as well, such as the Cheyenne mountain missile-tracking facility of NORAD. From the data of the current session, the Reptilian ETs appear also to have at least one facility hidden under a mountain. Mountain locations are ideal for the ETs as well as for human military personnel. They are rough terrain which cannot accommodate normal civilian traffic. It is not easy to farm, or to build cities or shopping malls on a mountain. Hidden bases under mountains have only to deal with occasional backpackers and other outdoors adventurists.

In the underground Reptilian facility that I observe in the current session, there are at least two types of subjects. The first type is the normal Reptilian ETs themselves. They apparently enjoy environments with dim lighting. The second type of subject seems to be a genetic hybrid, a cross between humans and Reptilians. These hybrid subjects are located in the same area as the purely Reptilian beings, and have some characteristics that are clearly more human than Reptilian. They enjoy brighter light in their surroundings, and they have some human facial features as well. Their consciousness also feels more human, although not totally so.

It is of interest to note that the hybrid subjects whom I observe in this session are fairly young, although they are not children. If the subjects are raised on Earth in this and other underground or hidden facilities, that would indicate that the Reptilians have been operating in our vicinity for a good number of years.

It is still unclear as to why the Reptilians are so concerned about secrecy relating to their activities. They are not the only group with underground facilities. Why is it that they desire to block our remote-viewing probes of their activities? The Greys also have been experimenting with hybrid genetics, but the

Greys do not now block humans from remote viewing them (although this was not always the case). Clearly we will need to know more about the activities of the Reptilians. Too many critical questions remain unanswered.

Chapter 18

THE REPTILIAN COMMAND CENTER

Based on the previous chapters on the Reptilian ET species, the following are apparently the case: (1) the Reptilian ET species exists; (2) it is currently involved in a significant military confrontation with another group, possibly the Greys; (3) Earth and humans have been indirectly caught up in this military confrontation; (4) the Reptilians have at least one underground facility on Earth in which both they and a Reptilian/human hybrid species work together; and (5) the Reptilians have some future plan for Earth and humans, although we do not yet know any specifics for this plan. For us to obtain a more complete picture of this situation, we need to identify and understand the center of operations for the Reptilians. Where is their primary base, their command center for all of their activities on and near Earth? What is the nature of the operations conducted at this command center? Why do humans not yet know anything about this command center? These are the questions that are addressed in the remote-viewing session for this chapter.

The essential cue and qualifiers for this chapter's target are as follows:

TARGET 4231/9902

Protocols used for this target: Enhanced SRV

The viewer perceives through the consciousness of the Galactic Federation

Contact Person for The Farsight Institute to remote view the command center for the Reptilian ET activities occurring on Earth (at the time of tasking). In addition to the relevant aspects of the general target as defined by the essential cue, the viewer perceives and describes the following target aspects:

- a description of the general layout and structure of the command center
- a description of the location of the command center
- the activities that take place in the command center
- the psychological, emotional, and intellectual aspects of those who work in the command center, as would be obtainable from a collective deep mind probe of those subjects
- the purpose of the command center

6 May 1998
11:28 a.m.
Atlanta, Georgia
Protocols: Enhanced SRV, Type 2
Target coordinates: 4231/9902

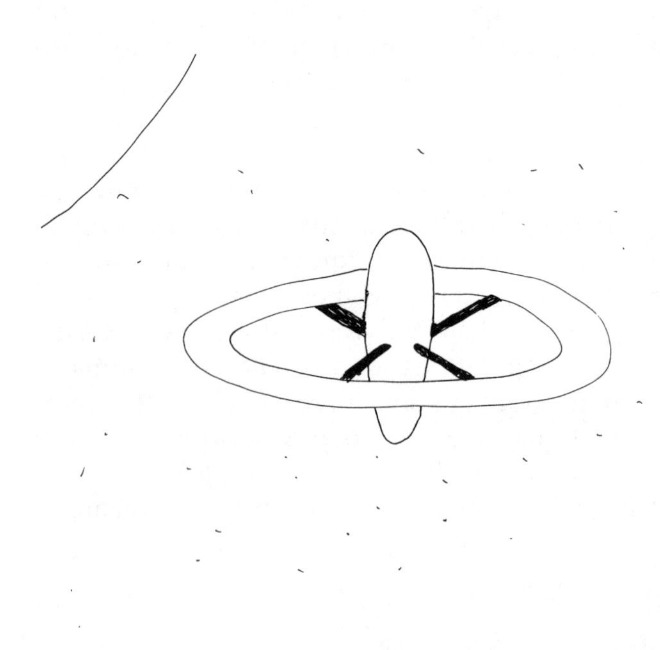

The session begins with a perception of a subspace being surrounded by light. The next impression is of a circular or curved artificial metallic structure. I note that the structure is surrounded by the color black with numerous pinpoint white dots, and I deduct a star field. I also deduct the idea of a mother ship. My next Phase 1 impressions are of a subspace subject, followed by a closer view of the artificial structure, and I deduct a space station. My Phase 3 sketch closely resembles a donut-shaped artificial structure in space near two planetary objects.

In Phase 4, I note the reflective and circular nature of the artificial object. Subspace beings have an observing interest in this target. The target seems associated with a plan followed by some groups. There is something large and circular in my field of view. It is like an orb or a planet. There is also a smaller circular object in the center of my field of view. The smaller object is highly reflective, bright, shiny. I am in space. It is cold and black.

I am also perceiving some humanoid subjects in a different location who are concentrating on something. They are planning something, or perhaps they are involved in a plan of some type. There is a room in the structure with technology. I am in the room now. There is technology all around, and I am deducting something related to the Mars96 Russian space probe. I sense that the subjects are concentrating on something "out there."

There are two locations within this target that are connected by activity or interest. I focus on the first place, and I perceive subjects within a land-based structure. It is a facility involved in controlling something. There is complex technology here, and I deduct a control room. I draw a sketch of the facility. My sketch resembles a tall, angular structure with both dish and straight antenna on or nearby the structure. Shifting my awareness to the second place, I perceive a structure in space that is curved, reflective, and spinning. It is a fast-moving object surrounded by a black void.

Moving to the next most important aspect of this target, I perceive nonhuman subjects, and I deduct Reptilians. The structure is circular and multi-level. There is a room in the structure that is somewhat plain with angled surfaces.

There is one subject in the room. He appears to be looking at me and aware of me. There seems to be a sense of disapproval of me, of what I am doing, or of how I am doing it. I execute a deep mind probe of the subject. He is resistant to the probe. He seems to be mentally "kicking" or fighting back, as if I am an invader. I search for the cause of the resistance, and I perceive the emotion of anger and the concepts of violation, control, secrecy, and security. I deduct the idea of a security oath, and I perceive the fear of superiors.

Still within the deep mind probe, I search for that which is being hidden from me. My awareness immediately shifts back to space. There is a distant structure, a highly advanced ET ship. It is hidden in some way, as though cloaked. I move my awareness to the ship, and I find it to be curved and stationery, but slowly spinning. It has lights on its exterior.

I move my awareness into the structure and perceive an advanced vehicle. I perceive the concept of something secretive. This is a ship, and there are subjects inside. I also perceive the mixed concepts of fear and resignation.

Some type of security alarm seems to have been activated. Everyone in the ship is standing still or moving very slowly. There is the sense that they cannot get me to leave, nor can they stop me from watching. But if they do nothing, I will perceive no activity and thus learn little.

I cue on the most important aspects that I am supposed to see related to this scene and the target cue. I immediately perceive a small cylindrical object surrounded by other shining metallic objects. It seems dangerous, like it will explode. It is associated with the concepts of "hidden," "secretive," "bomb," "explosive." I cue on the purpose of the object. It creates a small explosion. It makes a popping sound. It causes something to malfunction. Somehow the small object is induced to pop or explode, creating a malfunction. The device is altered or manipulated from a far distance. It appears to be from the ET ship. The ETs cause a weakening of the object in one location, and then the object pops.

I move to the location that will optimally reveal the reason for exploring this object. I am in space. There are stars all around. There is a very small object in the middle of an area of what appears to be small parts and pieces, debris. The small object is circular. It is functioning perfectly. It is hidden in the debris. It is set up so that no one will know or suspect it for anything but trash. But it works.

Moving closer to the small object, I perceive that it is angular, compact, highly advanced, and it is associated with something that is fraudulent. It has lights, computer circuitry, and a power source. It is a fraudulent thing somehow. It is a trick, "fool's gold." I cue on the purpose of the small object. It is a moderate-strength prototype/testing device—a weapon of some kind. But it is not useful against its intended target.

I move to a location where I can perceive the meaning of the intended target. In this new location, there are ET ships that are flying quickly. I now perceive that the small object is a "Star Wars" type of weapon, similar to that once publicly valued by the Pentagon and briefly supported in principle by the Reagan administration. But it does not work well.

Discussion

Following my own interpretation of these data, there is a Reptilian command center in space near Earth. It seems to be located somewhere between the Earth and the orbit of the Moon. This command center is cloaked in some fashion, invisible to normal human eyesight. It is probably also invisible to electronic surveillance of most types known to humans.

The Reptilians are apparently highly interested in maintaining the secrecy of their operations. They are now totally aware that we are capable of remote viewing their facilities and their activities, and they are adapting their behavior in a defensive fashion. Apparently they can do little but stand still when we remote view them.

This session pulls together a variety of seemingly separate themes. Apparently, there are humans somewhere who work out of a facility that can monitor some of the Reptilian activities. The Reptilians are involved in weapons manufacture. I think it is safe to assume that the Reptilians have the ability to manufacture effective weaponry against other ET ships. Thus, I interpret these data to suggest that the Reptilians are teaching some select humans to build prototypical weaponry. I suggest that these prototypes are designed to shoot down ships aligned with the Galactic Federation, particularly Grey ships.

The Reptilians are not really giving the humans effective technology. They are most likely seducing a small group of humans into thinking that they are obtaining useful weapons technology from the Reptilians. It is highly unlikely that the Reptilians would give humans truly effective weaponry. Only a fool would fail to recognize that humans would then have the capability of using this technology against the Reptilians themselves at a future date. Rather, what is going on here is a war of deception and intrigue.

I also suspect that some of the data in this session may relate to the Mars96 Russian space probe. It is possible that my subspace mind linked the activities of the Reptilian control center with the destruction of Mars96. Some of the data for this session suggest that a weapon has been hidden within a cloud of debris, and that the Reptilians used some small device to destroy something by making it seem as if there was a malfunction. These ideas fit both the known information relating to the Mars96 probe as well as the remote-viewing information.

This seems like a probable scenario that explains the mechanism by which the probe was destroyed, as well as at least one of the reasons for its destruction. From within the cloud of debris that once was a functioning space probe headed toward Mars emerges an advanced weapon, and a few military people are seduced into thinking that this weapon can help humanity win a war that the Reptilians themselves cannot win on the battlefield. The remainder of the world would continue to assume that the weapon is simply a piece of debris from a destroyed probe, while the most secret elements of human military forces would be free to use the new weapon to take shots at passing ET ships.

Chapter 19

The Reptilian Home World

Time and the extraterrestrials are a strange mix. Since ET ships can move through time with the ease with which we walk across a street, it makes little sense for us to target the Reptilian home world at the current time. The Reptilians may be coming here from a time long in our past, or even a time far in our future. What makes more sense is to target the Reptilian home world at the peak of their civilization. That way we can discern key characteristics of their nature as a civilization that may help us understand their current behavior. Since the Reptilians are here interfering with our home world, it only makes sense for us to shift our awareness to their home world. They have inadvertently gotten our attention. Let us now discover more about these uninvited "guests" by asking from where they came.

The essential cue and qualifiers for this chapter's target are as follows:

Target 3298/0219

Protocols used for this target: Enhanced SRV

The viewer perceives through the consciousness of the Galactic Federation Contact Person for The Farsight Institute, to remote view the home world of the Reptilian ET species that is currently operating on or near Earth (during the apex of their civilization on their home world). In addition to the relevant aspects of the general target as defined by the essential cue, the viewer perceives and describes the following target aspects:

- the physical environment of the subjects' living conditions
- the age and gender variations among the subjects
- the emotional state of the subjects
- the dominant groups among the subjects, including any governmental organizations
- the primary thoughts of the collective consciousness of the subjects
- the level of technology available to the subjects
- any environmental or other problems that affect large numbers of this species

27 April 1998
11:53 a.m.
Atlanta, Georgia
Protocols: Enhanced SRV, Type 2
Target coordinates: 3298/0219

Phase 1 begins with the repeated perceptions of hard, artificial structures. I deduct Reptilians on numerous occasions. I also deduct the idea of a gladiator. I perceive sentient subjects. After my fourth ideogram, I deduct the Reptilian home world, and I sketch a group of structures, deducting a village. I also perceive something in motion. Phase 2 data are particularly interesting for this session. I perceive a mixture of noises. Temperatures are very warm, and I deduct Africa. Colors are blue and green with bright luminescence. There are sweet tastes, and I deduct fruit. Something smells organic, like healthy soil. My viewer feeling is that the place is beautiful and that I like it. My Phase 3 sketch appears to represent a curved surface of a planet on which a collection of structures is located. Something is moving quickly over the structures. There is an orb in the sky, perhaps a sun or moon.

In Phase 4, I immediately perceive the taste of blood. Clearly something is both salty and red. I move to the optimal location with the widest useful angle perspective, and I perceive a planet with what appears to be a sun overhead. My attention is drawn toward a particular location on the planet. I am hanging in nothingness, a void. The planet is below. There are a multitude of lights on the surface of the planet. There is both land and water visible to me. I execute a movement exercise to gradually move to an optimal location to perceive the center of the target. As I move my awareness, I perceive that my vector of motion passes through turbulent atmospheric conditions. I am approaching land, and this place feels like the lush tropics. I have the deduction of the "time of the dinosaurs."

There is a great deal of vegetation here. I now perceive a single male subject who has a humanoid shape. But the subject does not feel human. He is old, and I have the deduction of a wise elder. There is a path cutting through the vegetation to a clearing. The subject does not appear to be perceiving me. Other subjects are nearby in the lush tropical vegetation. I perceive that this is a happy time, even between subspace and physical relations. It is a good place, a home world.

I probe the emotional collective deep mind of the target subjects, and I perceive them to have highly strung, strong emotions. I sense barking and yelping sounds, and I perceive the concept of fighting. There appears to be a collective personality that is genetically programmed. It is simultaneously reactive, sensitive, and combative.

I target the male subject that I perceived earlier, and I execute a deep mind probe. He is a leader, both older and wiser than most of the others. His clothing has multiple parts, in the sense that it is not a one-piece suit. I can discern that he is aware of the bigger picture. His skin appears dark and somewhat firm, yet it is also soft and silky, almost slippery, but dry.

I cue on the environment in the central target area, and while I find it to be lush, I do note that there is some pollution and mismanagement. The abuse of the local resources seem very typical of current human standards, but things are not yet as bad as they are on Earth today. Some of the environmental waste in this target location is very toxic. But I note an abundance of plant life that seems healthy.

Shifting my awareness to another essential aspect of this target, I perceive a fast-flying metallic object. It is curved in shape, and I perceive that it is predominantly used for atmospheric transport. Moving inside the flying object, I find it cramped with little empty space. It is densely packed with complex technological parts. I move my awareness to the aspect of this flying object that is the most important to observe, and I perceive a subject wearing a uniform. He is a pilot. He has a personality suited for command. This feels like some central place inside the object. There is technology here controlling something.

Once more shifting my awareness to the next most important aspect of this target, I again perceive the planet from a wide-angle perspective. The environment is generally good. The mismanagement of waste appears minimal. There is a rich atmosphere. In general, things seem optimal here. There is technology, but it is not super advanced. The place also does not feel overly populated, although there are many subjects here.

Since it seems clear that this location is not Earth, I execute a procedure in which I compare the current target location to Earth at the target time. When I probe the Earth at the target time, I get the sense of things that are out of control. Things are primitive and disruptive. Much is turbulent and chaotic. Earth is in a primitive state, like the planet is either coming apart or coming together. I perceive rocks, and I deduct volcanic activity.

I cue on life on Earth at the target time. The environment is not easy for life on Earth. Something has happened here. It is a burned planet. There is fire and rock, and the entire planet seems a cinder. All the building blocks of life are present, but it will take time to evolve. It is like life is supposed to evolve here again, but differently. There was an evolutionary mistake, and something had to change. It is like the planet has been wiped clean. It feels like this is Earth, but it also feels like this is an entirely different kind of Earth, an alternate Earth, or not my Earth.

Feeling somewhat confused with my perceptions, I decide to cue on Buddha, a subject with whom I have had some remote-viewing interactions previously (some reported in *Cosmic Voyage*). I quickly perceive him, and I execute a deep mind probe. I cue on the question of what is going on in this session, and I perceive the flavor of his thoughts.

This is an alternate time reality. It actually does exist. It is as real as myself. There is no "good" or "bad" reality, just a continuum of differentness. All subjects struggle for their own existence in a "slice" of experience, like a fish in a small area of a large body of water. Yet everything is connected, in the sense that all aquatic life could be located within the same body of water. There is nothing really "separate." There are degrees of distance between slices of experience, geographical, dimensional, time, etc. But there is only one existence.

I am experiencing a state of cross-dimensional contact. The Earth need not be "my" Earth, although the Earth experienced in the session does exist somewhere in some reality. The current target is significantly separate from Earth's current time stream as I am now experiencing it, and given current conditions and realities on Earth.

I thank Buddha and end the session.

Discussion

This session adds further complexity to my analysis of our current situation. The Reptilians visiting Earth in our current time frame may not even be from our own dimensional reality. If they have ships that can span time and space, then getting here from almost anywhere is probably possible.

It is of interest to ponder whether or not my perception of the Earth in the Reptilian's own dimensional reality represents a time early in Earth's history, or if it is an Earth that is in our future. Some of the data seem to suggest that life needed to start over again on Earth. This might indicate a situation in which life on Earth was destroyed, perhaps purposefully, so that a new order of things could emerge. I wonder if this is an Earth that the Reptilians want to create, an Earth in which all human life is eradicated so that a new form of Reptilian life can prosper. This thought seems too horrible to contemplate. On the other hand, perhaps the Earth that I perceived is Earth in the distant primordial past.

Yet one thing seems certain. The Reptilians evolved from an environmentally rich home world that was lush, warm, and tropical. In their most golden era, they were a happy, although apparently quick-tempered species. I do not know how long this golden era existed. I do not know what happened to their home world after the peak in their civilization. Did they eventually destroy their environment, or did they continue to maintain it? After they learned to travel through space, did they favorably colonize other planets, bringing a golden era to far-off places? Or did their own evolutionary history mirror that of humanity, full of ups and downs, golden eras followed by dark ages? One day human remote viewers will have the time and the resources to answer all of these questions.

The existence of a prior golden era for a civilization does not change our evaluation of the behavior of the Reptilians who reside in our own current time frame and reality. While we now know a measured amount regarding their history and their planet of origin, we still need to know more about Reptilians as a group today. We need to know about their society, their current civilization. We need to know more about how "our" Reptilians live among themselves, and how they organize themselves. I address these issues in the next chapter.

Chapter 20

Reptilian Society and the Renegades

It is one thing to observe the activities of a species, and quite another to obtain a complete overview of their civilization. The advantage of using the Social and Political Protocols to explore Reptilian society is that one can learn more in one session using these protocols than is possible from many sessions using Basic or Enhanced SRV. From the previously presented chapters involving the Reptilian ETs, they appear to be a confrontational species. We need to know more about the groups in their society. Do all Reptilians support the overall Reptilian agenda involving humans and Earth? What are the characteristics of the most dominant group in the Reptilian government? Are the Reptilians who are interacting with humans on Earth working as an extension of their official government, or are they a renegade group? Answers to all these questions, and more, are vital to our understanding of these beings.

The essential cue and qualifiers for this chapter's target are as follows:

Target 9023/4792

Protocols used for this target: SPP

The viewer perceives through the consciousness of the Galactic Federation Contact Person for The Farsight Institute, to remote view the civilization of the Reptilian ET species that is currently operating on or near Earth (at the time of tasking). In addition to the relevant aspects of the general target as defined by the essential cue, the viewer perceives and describes the following target aspects:

- the overall civilization as it exists in the current tasking time frame
- the most dominant group in the government of the civilization
- the group that is physically located on or near Earth in the current tasking time frame
- any group of the Reptilian ET species in the current tasking time frame that is opposed to the current activities of the primary group of Reptilian ETs that is operating on or near Earth
- the primary thoughts of the collective consciousness of the Reptilian ETs who are interacting with human governments (or their employees) on Earth

30 April 1998
12:10 p.m.
Atlanta, Georgia
Protocols: SPP, Type 2
Target coordinates: 9023/4792

In Phase 1, I identify a target macro with three significant groups. The groups are separated by the way they think and perceive. The macro itself feels cohesive despite sub-macro variations among the groups. One of the groups is a central group that seems to have authority over the macro society. There is another sub-macro group that is held together by a sense of cultural commonality. I perceive that it is internally homogeneous. I also perceive that there is a loosely assembled populace, or perhaps a collection of groups within the macro-society. This collection of groups has a common need that is satisfied by the larger society.

In Phase 2, I sense that the macro-society is huge. There is a large subspace component to this society, as well as a very large level of variation among the beings on the subspace side. It is a diverse collection of groups. There is a central pull that binds these groups together. The pull feels weak, in the sense that a group could resist and defy it if it chose to do so. But there appears to be no useful competitive alternative attractor for this society with a stronger pull. It is as if the central authority in this society is the only game in town worth playing. The pull toward the center is constant, like the steady gravity of a planet or sun. I perceive no sense of rigid authority or any enforcement of allegiance to the central authority in the macro-society, however. Most of the activity on which members of this society are focused reside in the subspace realm. There is also an extensive use of telepathy among these subjects. The physical side of this society seems highly dependent upon the subspace authorities.

When I probe the significant leader of the macro-society, I perceive no resistance to my probes. There seems to be a benign welcoming. There is no sense of brutal authoritarianism in the subject's mind. Yet the consciousness of the subject conveys the sense of powerful leadership. This being has a large and complex array of means to assert authority. The relationship of this subject to the sub-macro groups is clear. This subject's job is to connect the disparate groups to the center of the macro-society.

In the Phase 2TM consciousness map, I perceive that the physical subjects of the society are being led, somewhat blindly. On the subspace side the focus is not on control but on collective advancement. In the Phase 3TM sketch of the target macro, I again identify three groups, and I label them in the conventional fashion, G1, G2, and G3.

G1

G1 is a medium-sized group with both physical and subspace aspects. There is a highly focused concentration of authority in this group. There is only a single authority attractor in this homogeneous group. The ideology of this group contains a theme of allegiance to authority. The pervasiveness of this theme is both strong and total. There is rigidity in the political ideas among the membership of this group, and there is a notable lack of deviation or variation in this regard. Membership in this group feels almost bureaucratic, and I feel no sense of emotional competition between the members of the group.

There is a significant level of physical activity among members of this group; however, they use telepathy extensively in communication. The subspace side of this group dominates the physical side. I again perceive a highly rigid central authority for this group, and the macro-society accepts this authority.

When I probe the significant leader of this group, I perceive a subspace being who is a central authority figure. There is a softness to the consciousness of this being. He is male. I sense that he is aware of my probes and not resistant to them.

Shifting my awareness to a typical non-leader member of this group, I perceive a gentle consciousness that is non-aggressive. This subject is not emotionally demanding of the central authority. The subject has unquestioning obedience to the central authority, however.

G2

Moving to the next identified group (G2) in the society, I perceive a large population. There is significant physical genetic variation within this group, but minimal subspace variation in development. This group is highly different from the first group (G1). The organization of this group is predominantly political. There is the collective sense that the group needs to compete with other groups in the larger macro-society as a means of survival and advancement. The leadership is judged by how well it performs this role of guiding the rather crude struggles with the group's neighbors. The group is highly competitive and potentially hostile in a general sense.

The activities of this group appear more focused on the subspace realm, and there is extensive use of telepathy as well. However, the relationship between the subspace and physical sides of this group are not highly supportive. The subspace side has only a weak influence over the physical side, and the physical side is quite disappointed in the quality of the subspace leadership.

This group is driven by competition and petty jealousy. It is not in total harmony with the center of authority in the macro-society. There is a lack of understanding regarding the motivations of the macro-authority and culture. The macro-society looks at this group with significant dismay and worry. It is like the macro-society does not know what to do with this group.

Shifting my awareness to a significant leader of this group, I immediately perceive that this being is a Reptilian. He is a male, and he is a fighter. This subject is resistant to my probes, and fiercely so. It is like he is being held down and forced to let me perceive him. He is the political head of his group. He is in power due to his own abilities to compete politically. His position is not totally secure. If the group becomes disappointed with his leadership, he could be ousted.

Shifting my awareness again, I focus on a typical non-leader member of the group. Again I find a fighter. There is the sense of a need to seek prey. This subject's perception of the collective good includes help from the group leadership to hunt, compete, struggle, and persevere. The subject questions leadership only if group performance falters with respect to competitive struggles.

With a consciousness map, I determine that the emotions of the physical side of this group are like that of a typical football game, excited and highly strung. On the subspace side there is sensitivity and panic.

G3

The final group, G3, has a large population. There are moderate physical variations within this group, and subtle differences among subspace types. G3 has a group mentality. The members define their own existence in terms of their participation in group activities. This group participates in the activities of the larger macro-society because it is in the nature of all members of the group to struggle and advance in cooperation with other groups, or at least to work within a group setting. This group could and would struggle alone, but that is not its preferred way of advancement, or of living. There appears to be no single authority or individual ruling this group. It feels almost like there is a collective means of resolving disputes and problems.

This group is highly focused in its activities on the subspace realm. There is also close cooperation between the subspace and physical sides. The macro-society values this group. It is a stable and reliable component of the larger society.

From a consciousness map of this group, I perceive subspace emotions of worry relating to the behavior of the larger macro-society. The membership of this group perceives participation in group activities to be almost spiritual in nature. Since the macro-society has a wider and more heterogeneous group membership than this group, the members worry about the reliability of the larger society to satisfy their needs. But there appears to be no alternative for this group than to participate within the framework of the larger macro-society.

The Macro-Society Developmental Trajectory

The development trajectory of this macro-society indicates an early period in which chaos and group fragmentation dominated. There is a "Wild West" flavor to this early time. Many groups were at odds with each other.

Soon after this chaotic beginning, I perceive a crucial point of transition. I sense a determination among the members of the society to resolve long-term potential for chaos, and to change the macro-society to operate in a more orderly direction.

Near the end of the identified developmental trajectory, there is greater coherence in the society. This society has learned how to resolve many of its inner conflicts. There is smoother social functioning and more isolated potential for conflict, especially violent conflict.

Discussion

All in all, the Reptilian society seems rather normal when compared with human standards. There seem to be normal variations among various groups in the society, and I perceive no overall sense of strong authoritarianism. However, the second group examined (G2) seems remarkably different from the rest of Reptilian society, and it is probably worthwhile to comment on this group specifically.

Note that only one of the qualifiers of the target cue focuses on the group that is physically located on or near Earth in the current tasking time frame. All other qualifiers focus on either the overall Reptilian society, the dominant group in that society, or any group of Reptilians that may be opposed to the activities of the Reptilians operating on or near Earth. Since my data collected in other sessions suggest that the Reptilians interacting with humans are quite authoritarian and potentially hostile, those Reptilians are likely associated with group G2. All of the other elements of Reptilian society seem relatively benign politically. There seem to be no inherent traits among Reptilians generally that would make me believe that their entire species is hostile. However, the observed behavior of Reptilians interacting with humans closely matches the characteristics of group G2. Moreover, the deep mind probes of the leadership of G2 qualitatively parallels that which I experienced in other sessions with this group.

My interpretation of these data now lead me to suspect that group G2 is a renegade Reptilian faction. This group is qualitatively different from the remainder of Reptilian society. It is similar to human society, as we too have our authoritarian or totalitarian factions. These factions have often behaved in total disregard to the wishes of the larger community. It now seems clear that there exists a group of Reptilians who have a chip on their shoulder. They are highly competitive and emotionally arrogant. I suspect it is this group that is at war at the current time, not the entirety of Reptilian society.

If my interpretation of these data are correct, then we truly have a significant problem. If the larger Reptilian society itself cannot control this group, then how are we to control them? If this group defies the authority of its own larger species, then will it obey a directive from the United Nations? I suspect that this Reptilian group will do whatever it wishes to ignore, manipulate, or circumvent human authorities. In my view, it will not be possible to interact openly and productively with this group of Reptilians, and it would be foolish for us to try to do so. Our own experience with human totalitarian and authoritarian regimes should help us to understand the dangers associated with working with such groups. Based on my observations to date, we will need outside help. Unless further data suggest another course of action, I see no alternative to this. As relative newcomers in these galactic woods, we cannot, we must not, act alone.

PART IV

The Greys and the Galactic Federation

Chapter 21

The ET Technology Objects

During the first half of 1997, I obtained information from an unusual but seemingly reliable source that suggested that the Grey ETs might be placing technological objects in a desert location, possibly in the northern United States. The information suggested that the objects may be spherical in shape, and that the U.S. government (or at least part of the government) knew about their location. At first I did not think much of this information. I had no way to corroborate it, nor did I understand how it might involve my own activities. If the Greys were placing objects in chosen locations, so be it! At least that is how I felt about this matter, until I was given the following target by a colleague for one of my regular solo viewings. After May 8, I took the matter of strategically placed ET technology much more seriously.

The target cue for this session employed an older version of the Institute's cuing formulas. The Type 3 target was given to me as one of my regular practice sessions by someone at the Institute. The cue is as follows:

The Grey ET technological objects in North American Desert (at the moment of tasking)
8 May 1997
11:30 a.m.
Atlanta, Georgia

Protocols: Basic SRV, Type 3
Target coordinates: 0476/9827

Using Basic SRV, I perceive land and a structure. In Phase 2, I become aware of the sounds of machinery, as well as a whirring noise. The textures are glossy and polished. The temperatures are warm. I perceive the colors blue, black, silver, and gold. The level of luminescence is low, and the contrasts vary from medium to high. I perceive salty and bitter tastes, and I smell a variety of scents. The magnitudes of the dimensions for the target include intense energetics that are somehow compact. I deduct a computer chip. My Phase 3 sketch is of a rectangular structure on which is placed another smaller rectangular structure. The smaller rectangular structure has numerous wavy lines within it.

In Phase 4 I perceive bright light. There is energy at the target site, and this energy is a mixture of a variety of energetics. I perceive particles associated with this energy that are flowing in orchestrated or uniform directions. There is the feeling of pathways along which this energy is channeled. Something at the target site is both large and small. It is big on the outside but small on the inside. Again, these are intuitive feelings, and they may not make logical sense at first. It is as if activity is occurring in the inside of something that is on a small scale. But this small scale activity results in something else happening on a large scale.

I continue to perceive bright light as well as the sense of flowing energy. I am also beginning to perceive emotions at the target site. These are not human emotions. The emotional quality is very uniform, as if the feelings are smooth and flowing throughout a collective consciousness. I perceive a hive mentality, and I deduct the Greys. The emotions of the subjects feel both wide and deep. They are emotions from a collective consciousness that penetrate. I begin to perceive the beings physically. I note their eyes and skin. These beings appear to be nonhuman, but they are humanoid. Their consciousness and physicality extend deeply into the subspace realm. I have the guided deduction of the Grey home world, and the deduction of the so-called Hale-Bopp Companion. (Note: The "Hale-Bopp Companion" refers to what appears to have been some type of dimensional gateway or portal that was briefly located near the comet Hale-Bopp and which has been repeatedly viewed by many remote viewers at The Farsight Institute.)

There is a structure associated with these beings. It has technology, and I perceive it is constructed of metal (at least in part). The structure is small, and I sense that the emotions of the associated beings are deeply connected with this object. It is like there is a collective consciousness of beings that extends perceptually throughout time and space. There is very deep awareness that is associated with this target. Inwardly, I reflect that this seems like consciousness as an art form. I do not perceive any human emotions associated with this aspect of the target. I deduct zillions of Greys. Humans do not have consciousness like this. It is clear and beautifully deep.

There are also humans at the target site. They do not seem as "rough" as regular humans. But they are certainly different from the Greys. I probe my Phase 3 sketch and perceive technology. The structure in the sketch feels metallic and non-metallic simultaneously. It is definitely a structure that contains a high level of technology. I deduct molecular-level technology. It is as if there are circuits inside solid metal, and the metal is alive with a network of channels containing the flow of energy. The object acts to control something. The technology involved with this structure may be a hybrid biological technology, and perhaps on a very small scale. It is as if this object is alive.

Within the object, something small and possibly biological seems almost alive and moving in a purposeful way. It is not purposeful solely for the organism, but also for a larger purpose that is unknown to the organism. The target structure has microcircuits, or pathways, with great internal complexity. The thing is almost self-aware. But it is a device nonetheless. It seems to have been designed by beings who wanted all things, even their creations, to join them in self-awareness. The internal mechanisms of the target structure seem to be arranged in a stratified pattern, with open areas below the structure and densely packed technology on the upper level.

I begin to perceive that some activity is associated with this target structure. The activity is orchestrated by many beings, apparently not human. Something is happening quickly. I am now aware of a bright light in the sky. The concepts associated with that light are "watching," "pulsing," "rhythmic," "movement," and "orchestrated." The sense of deep penetrating emotions are pervasive with this target. There are multiple structures moving in the air, and I deduct ET ships. There is a bright yellow light associated with a target, and it appears to be in the sky. There are intense levels of energy associated with these airborne structures.

I am now becoming aware of a large subspace presence associated with this target. It is like the physical stuff originates from a small hole in time and space that leads to an enormous subspace existence for some species. Allowing my own consciousness to extend through this hole to the other side, I perceive vast numbers of beings, and a collective consciousness that seems to extend without end. On the subspace side of this target, there are emotions that are mixed across what feels like different levels of Being.

Discussion

Remote viewing targets related to Grey ETs are always a remarkable experience for me. The aspect of their lives that stands out is the quality of their consciousness. The only way I can describe this quality is through an analogy that does not do it justice, but that will have to suffice. Whenever I extend my consciousness into their own, I get the sense of an endless clear body of water. It is as if one could drop a pin and hear it across the universe due to the seamless clarity of their mentality that extends, apparently, everywhere. The consciousness of the Greys is always collective. They are individuals, but their minds are linked in a way that makes the totality of their collective seem as though they are almost one being. In my opinion, they have raised consciousness to an art form, and their ongoing awareness is nothing less than breathtakingly beautiful. This is not to say that they do not have problems, or that humans should try to follow their evolutionary path of collective existence. Nonetheless, just as I can recognize the beauty of a flower when I see one, I can similarly recognize that the quality of their collective consciousness is sublime.

The current session lends support to the idea that there are ET technological devices strategically located in the United States. I have remote viewed these devices more than once. Apparently, the Greys use "consciousness-enhanced technology," a phrase I once heard used by Stephen Greer, a gifted and courageous medical doctor who has been involved in efforts to encourage Congress and the president to openly investigate the extraterrestrial phenomenon. The technological objects that are the focus of the current session are apparently good examples of this type of consciousness-related technology. Indeed, it appears that the technology of the Greys is sufficiently evolved that at least some of their machines are approaching the level of being self-aware.

After reflecting on this session, I began to wonder how humans would react to the idea that consciousness can be more than a metaphor, and that the members of another species consider their own development in consciousness to be the primary goal of their collective evolution. Will humans continue to focus their attention on the physical aspect of life, marginalizing spiritual concerns into a soon forgotten once-a-week trip to a house of worship? It seems that we could learn so much from the Greys about their experiences in consciousness. Will we ever lose our fear of them, thereby allowing us to interact with them productively?

In my own mind, I feel certain that humans should not direct their collective evolution in the manner of that experienced by the Greys. We must not lose our individuality as we grow in spirit. But at least some of what the Greys have accomplished on the level of consciousness needs to be examined seriously by humans. The magnetic draw toward the beauty of their thought is nearly overwhelming to me. Such grace is truly Godly. Perhaps we can grow to experience that same penetrating quality of clear transcendental thought without sacrificing the individuality of our personalities.

Chapter 22

A Companion No Longer

In late 1996, The Farsight Institute published some remote-viewing data on our website relating to what appeared to be an interdimensional portal or gateway that was located (at least for a short time) in our solar system near the comet Hale-Bopp. This was soon popularly labeled the "Hale-Bopp Companion." After we announced our original remote-viewing results, a cottage industry seemed to form across the planet. Many people directed their attention toward the heavens in the direction of the then famous comet. They wanted to see the Companion. They looked across the Internet, sometimes with success, for previously posted photos of the comet that contained some potentially anomalous aspects. Amateur astronomers were the most active in the search for the Hale-Bopp Companion. No one to my knowledge ever claimed that the Companion was always visible near the comet. Some speculation suggested that it might be able to appear or disappear at will, turning itself on and off like the headlights of a car. As the weeks dragged on with no resolution to the matter, the comet disappeared on the other side of the sun.

Some remote-viewing Type 4 data collected by advanced students at the Institute during this period suggest that the Companion would change direction once the comet went behind the sun. Personally, I do not know what happened. I do not even know if the Companion was even following the comet, or whether it might have been located near the comet only briefly.

Readers should remember when they read this session that we do not know much about the Companion. If it was a dimensional portal, or something similar, its original size may not correlate with its current size at all. If the Companion is a product of advanced ET technology, we should abandon all preconceptions of what it is, or even what its purpose may be.

The cue for this session was written using an older form of tasking structure. I was given this Type 3 target "out-of-the-blue" by someone at the Institute. The exact wording of the cue is as follows.

> The object formerly known as the Hale-Bopp Companion (at the moment of tasking): The viewer will clearly perceive and describe the Hale-Bopp Companion. The viewer will clearly describe the subjects associated with the Hale-Bopp Companion. The viewer will clearly describe the location of the Hale-Bopp Companion. THE VIEWER WILL PERCEIVE ONLY THE INTENDED TARGET THAT IS CURRENTLY ASSOCIATED WITH THE ASSIGNED TARGET COORDINATES. THE VIEWER WILL NOT DESCRIBE ANY BEING, OBJECT, OR INTANGIBLE THAT DOES NOT EXIST IN THIS TARGET. THE VIEWER WILL REMAIN FREE FROM ALL NON-TARGET INFLUENCES.

15 January 1998
2:58 p.m.
Atlanta, Georgia
Protocols: Enhanced SRV, Type 3
Target coordinates: 1883/1129

My first ideogram feels soft and natural. I declare the target aspect a subject. I perceive the color beige and feel something soft. I sketch a humanoid's face. The second ideogram is again soft and natural. I again declare this aspect a subject, and I perceive something semi-hard and angled. I sketch a standing humanoid leaning over a rectangular object. The third ideogram is again soft and natural, and I again declare this aspect as a subject. I perceive something angular, both hard and soft, and the sense of something multiple. I sketch one humanoid sitting on something squarish while leaning over something else that is rectangular. Two other humanoids are facing this first humanoid while standing on the opposite side of the rectangular object.

The fourth ideogram feels hard and artificial. I declare it a structure, and I perceive the colors gray, blue, and yellow. Something is bright, shiny, and glowing, with high contrasts. There are energetics here, and I sense something humming. It is round, and I deduct an ET ship. I sketch a circle. My final ideogram again feels hard and artificial, and I declare it as a structure. I perceive something light and hollow, round and glowing, yellow and gold, and I deduct the idea of ET spheres. I sketch a circle with shading on the left side.

Moving to Phase 2, I perceive the sounds of wind. The dominant texture is polished. Temperatures are cold. Among the visuals, I perceive the color gold, luminescence that is glowing, and contrasts that are high. I smell manure, and something that smells like solder paste. The magnitude of the dimensions of the target site are short verticals, narrow horizontals, something sloping, something curved and rounded, something that is light, and all this associated with energetics of some sort. My Phase 3 sketch is of a sphere or globe. The globe is shaded on the left side.

In Phase 4, I perceive bright light that is glowing with a gold color. I again sketch a globe or sphere with shading on the left side. I perceive the concept of something round. There are motions at the target site that I describe as having "high energy." There are multiple objects that I perceive as having some specific purpose. I have the guided deduction of globes. The color is yellow, the temperature is cold. The object is very light, almost weightless.

I perceive what appears to be a different rounded structure on flat land. I deduct the idea of a dome and draw a picture of a dome on flat land. There is dirt, soil, ice, and melting water in a cold environment. Again I sense the color gold associated with a round structure, and I again deduct the idea of a dome. Something is hollow and round at the target site, and I again draw the sketch of what appears to be a dome on flat land.

I perceive a dome or curved structure on top of or embedded into the ground. It is hollow inside. There are subjects inside the structure. The outside climate is cold and harsh. There is land below the structure. The structure itself has thin walls, and it is made of artificial materials that are multi-layered.

Cuing myself to move into the structure, I perceive curved walls. There are multiple subjects, both male and female. However, there appear to be more males than females. I execute a binary technique that allows me to determine the types of emotions of the subjects at the target site. I find their emotions to be generally positive. I perceive people working at what appear to be desks arranged in a geometrical pattern, and I sketch this. I get the sense that these are offices of some type, and I sketch a wide-angle perspective of the arrangement of the desks or workstations. They are arranged in concentric curved rows.

I note that the subjects are wearing clothes, and that the environment is neat and clean. Some of the subjects are wearing uniforms. I deduct the sense of something military, as well as the idea of a satellite tracking station. The subjects are definitely humanoid, and I get the sense that this is a facility. It is complex. The emotions within this facility convey the idea of tension and worry. This is a working environment, and I have the guided deduction that this is a military facility.

Cuing myself to move around inside the structure, I perceive walls, partitions, and compartments. There is a complicated, maze-like layout within the structure, with lots of compartments. I deduct the idea of office dividers. The site feels like a main office. I closely examine one male subject who is wearing a uniform. I get the concept that this person has a high rank, like that of general or commander. I enter the person's mind with a deep mind probe, and I feel worry, concern, and fear. There is a sense of patriotism within this man's mind. But something is out of control. Things are not right. This target subject is very worried. It feels like he is worried about the future. There is some element of depression within his mind, as if he is in a deep funk. He does not know if what is happening is good, or if it could be controlled. He feels sad, almost weepy. It feels as if he knows things others do not, and that they are happier than him as a result. He feels some sense of responsibility and seems to be worried about failing.

I cue myself to move 1,000 feet over the structure. It is curved and rounded. It is on land that is both wet and dry. There are nearby rectangular structures. Moving to the center of the target, I perceive subjects wearing clothes. They're working on some activity, some project. The focus of this target seems to be purposeful activity associated with this project. When I cue on the purpose of the activity, I get the sense of deception, or sleight of hand.

I execute some binary procedures that help me focus on the designed purpose of the target. I find that the focus is not people or activity, but a physical description. I then move to the location that will be optimal for me to achieve the purpose of the target. I perceive a round, spherical structure. I draw a sketch of a globe that is shaded on the left side. I perceive other objects. I cue on the objects and perceive the concept that they are gold in color, round, and of ET origin. There is advanced solid-state technology inside the objects. I perceive no moving parts. I cue on the connection between the dome-like structure and the objects. The concept of being in control is very clear to me. I deduct the idea of "explosive," and I remember perceiving some desire to explode the objects. There is some connection between the dome and the objects, but it may not be symmetric. It is not clear if the dome people have any control over the objects.

I cue myself to move to another aspect of the target location. I find myself outside in a cold and windy, harsh climate. There is land with structures that are modern and rectangular nearby. The buildings are gray, and there are subjects who are human or are very human-like. I deduct a campus, a city, or a facility.

Discussion

Readers should understand that when the so-called Hale-Bopp Companion was originally targeted by viewers at The Farsight Institute, there was no clear indication of its size. The assumption was popularly made that the Companion was extremely large, possibly much larger than the planet Earth. But the initial remote-viewing data clearly indicated that the object was more than just a physical device. It seemed to be a dimensional window or portal of extraterrestrial origin, and more, much of which we could not understand.

I have always assumed that what was originally perceived could vary dramatically in size. A dimensional window or portal could expand or shrink as needed, given the circumstances at the time. It could perhaps shrink to a molecular size, or expand potentially to a size greater than that of our solar system. It is simply impossible to know the fixed dimensions of such advanced ET technology.

The object known as the Hale-Bopp Companion is clearly not a companion of a comet. Yet I do not know where it is currently. These data suggest that a governmental facility with a large domed structure is monitoring it, or perhaps searching for it. The humans in the facility are apparently very aware of the ET object. It is spherical, shiny, and has a golden color. The governmental facility is apparently in an area that has a harsh and cold climate, most likely in the northern or northwestern parts of the United States. Perhaps the Dakotas might fit this description.

The high-ranking personnel who are in charge of monitoring this object appear to be quite worried about the implications of its existence. Some of these personnel would apparently like simply to destroy the object. There is a tension in the air as they continue to observe it or to search for it.

I do not know the purpose of this object. It obviously can move through space, and it possibly can be used to transport beings and physical equipment. The object itself may be any size. I simply do not know its current size, nor can I discern its size from these data.

This object has caused more controversy than any previous UFO sighting except the famous Roswell crash. The existence of this object is probably extremely important in the drama that grips humanity today. Its existence is also evidence that some ETs want to force us into a psychological crisis of awareness. They do not appear to want to hide their ships or their controversial technology. They want us to see them.

Chapter 23

The Grey Society

The Greys have a very interesting society. With this species, we are talking about *very* large populations. Since the Greys are not limited to a single planet, their numbers are not limited either. Also, they have the technological capabilities to satisfy their material needs. Interestingly, however, though they lack for nothing physically, it is spiritual growth that they seek most strongly. Never have I encountered a species in which the quest to evolve toward God is felt as deeply. In my view, this species looks toward spirituality like a starving man looks at an apple. Thus, when we examine this huge society, it is useful to note variations in their organization, as well as in their spiritual outlook. Be sure to closely examine the qualifiers in the target cue. The cue is designed to identify groups of Greys based on their level of evolution. This is one society in which such a differentiation between groups truly makes sense.

The essential cue and qualifiers for this chapter's target are as follows:

Target 7984/2324

Protocols used for this target: SPP

The Grey Society (at the time of tasking). In addition to the relevant aspects of the general target as defined by the essential cue, the viewer perceives and describes the following target aspects:

- the macro-Grey society that is directly or indirectly involved with contact with Earth humans
- the most highly evolved group of Greys that is directly or indirectly involved with contact with Earth humans
- the least highly evolved group of Greys that is directly or indirectly involved with contact with Earth humans

16 May 1998
4:04 p.m.
Atlanta, Georgia
Protocols: SPP, Type 2
Target coordinates: 7984/2324

My first ideogram reflects a large target macro with intricate group heterogeneity. There is great complexity here, with many small units. I perceive that most of this population are subspace beings. The second ideogram represents a smaller sub-macro group that is homogeneous. It has particularly clear cultural definition. These are physical subjects. The third ideogram reflects a combination of physical and subspace subjects, and I deduct the Greys. There is some heterogeneity in this sub-macro group, and I deduct the idea of "shades of gray." My fourth ideogram reflects physical subjects who have a clear cultural definition. They are experientially connected in some way.

In Phase 2TM, I perceive a very large macro aggregation. The population numbers are simply tremendous. There is an extreme variation in physical and subspace types, a complete continuum. The target macro is very loosely organized, leaning toward anarchy. The dominant similarity of thought among the various subgroups is of an experiential sort, as if everyone experiences something different, and it is having the experience that matters. This society is highly fragmented with low levels of habitual behavior.

Most of the activity and communication in the macro-society is subspace-oriented and telepathic. The subspace side has a weak administrative role over the physical side. Variation among the sub-macro groups is tolerated, even when such groups are confrontational. The leadership of the macro-society reflects this low-key approach toward subgroups as well.

The subspace collective psychology of this macro-society includes the ideas of administration and governance. When the subjects are physical in this society, they find it hard to follow some of the subspace administration. Some of the sub-macro groups in the society are confrontational, while others are cooperative. There is a complete mixture of everything in this society. Yet the macro-society tolerates this diversity religiously.

When I focus my awareness on the significant leader of the macro-society, I perceive a subspace being who has total telepathic awareness. It feels as though my consciousness has difficulty understanding the nature of this subject's consciousness. The leader is as far as one can be from authoritarian. Indeed, this leader significantly expands the definition of tolerance in the direction of anarchy.

In Phase 3TM, I identify three sub-macro groups. I label them conventionally as G1, G2, and G3.

G1

G1 is a strange group. (That is a viewer feeling!) It is small in size, with a high concentration of authority. The type of authority seems highly sensitive to maintaining a flexible and egalitarian quality. There is almost a religious flavor to this group, held together by some kind of experiential/belief mixture.

There is little or no fragmentation with this group. The group's primary focus is on subspace activities, and the group itself is apparently a subspace group. The macro-society is not clear about what this group actually does, or how much it can or will do. The leadership is tolerant and compassionate, and the membership is similarly high-ranking in some sense.

More specifically, the group is highly homogeneous. If one can say that this group has a common ideology, the best word to describe it would be religious (although this is not quite right). This ideology is totally pervasive within this group. Politically, the group has an administrative flavor, as if confrontation is not part of their normal political nature.

Focusing on a significant leader of this group, I perceive a subspace subject who is caring and compassionate. This subject has generally positive feelings that are laced with some concern or worry. This subject holds an appointed position that must be approved by the larger group. In this sense, one may also think of this subject as a democratic leader, although a human understanding of democracy would probably not fit in this situation.

The collective psychology of this group contains a mentality that is a blend of philosophy and emotions. But the emotions are difficult for me to understand completely.

G2

G2 feels like a predominantly physically focused group. It is of moderate size. There is a high concentration of authority, in the sense that there is not much competition among individual members of this group. But the authority is irregular, with competitive or alternative authorities existing on a religious level.

There are political and religious themes that run throughout this group. High levels of fragmentation exist at this level. The group is aware of the physical/subspace interaction, and the subspace activity of this group seems to be coordinated with the physical activity.

This group is highly dependent on outside assistance or intervention of some type. The macro-society acts to protect this group in the same way that humans protect endangered species. The secular leadership is administrative in nature, and the group members rely heavily on this leadership. Indeed, the group membership seems vulnerable and dependent in some fashion.

The significant leader of this group is male. He has a political mentality. He has good administrative capabilities and holds comparable responsibilities as a project coordinator.

G3

The final group upon which I focus is very large with primarily a subspace focus. There is also a great deal of physical variation among the members of this group. The group appears to use telepathy for all of its inter-group communication.

In some way, this is a simple culture, and other ways it is highly complex. The concentration of authority is quite diffuse, with multiple attractors within the group. The pervasive ideology is nearly religious, although this description does not fit exactly. There is the sense that work and membership provide service, and that service is the political orientation or belief structure of the group. That is, the group believes in service in the same way that humans might believe in patriotism.

The subspace side of this group views the physical side almost in an artistic or even experimentalist fashion. The group seems somewhat isolated from the remainder of the macro-society in some way, and the macro-society does not entirely understand the group.

Leadership of this group is collective. While I do not understand the mechanics of governance within this group, it is clear that there is no one single leader. The group members feel very secure in their membership, almost as if they feel that they are members of a family.

The Macro-Society Developmental Trajectory

I identify three significant points in the development of this macro-society. In the beginning there was a significant crisis that led to a complete disruption of the society. There was a war, and various groups disintegrated.

At the end of the trajectory, I perceive lower emotional energy, even numbness relative to that found at the beginning of the trajectory. There is calm and peace at this end point.

The period between the beginning and the end of this trajectory feels smooth, as if there is a uniform transition between the two end points. But there is one critical period, a point of breakthrough in the awareness of this society. It is a critical or pivotal point. It is like a new element is added at this midway point that causes an evolutionary shift.

Discussion

This is such an interesting society. The Greys are so different from us. Their consciousness has an innocence that is almost childlike. They are capable of witnessing all human activity, and they have a rich history of experience that would fill any galactic library. Yet they have turned away from simple physical enrichment. As a spiritual species, or at least a species dedicated to the inner growth of that which is spiritual, is how I perceive these beings.

I am always surprised when I hear someone say they fear the Greys. I certainly perceive no element of militaristic hostility, authoritarianism, or aggressive emotions from any group of Greys. In my view, fear of this species makes no sense. They themselves do not feel fear, at least not the way we feel it. When they perceive our fear of them, it must baffle them. They have much to tell us, much to teach us. In my view nothing is so important for humanity than for us to learn how to seek God more clearly, more scientifically, and more lovingly. Too often we speak of God's love in the context of a fear of punishment. Religiosity and spirituality are two separate things. My experiences with remote viewing suggest to me that our understanding of spirituality, and our practice of religion, would be enhanced significantly by studying how the Greys perceive and interact with the heavenly realm.

We should learn what the Greys have learned, and we should add their knowledge to our own, thereby enhancing our own evolutionary growth. My experiences suggest that the Greys do not wish to hide anything of themselves from us. They are an open book. What is required is courage on our side to open our eyes and minds and avoid the blinding influence of fear, and to accept a gift of love that they seem so willing to share.

Chapter 24

Galactic Federation Headquarters

If you place 15 people in a room for an afternoon, within a short time someone will collect money for a pizza run. Wherever there are sentient beings, they organize themselves. This is true not only of humans. Organization is as natural as existence itself. Nearly every form of life that I know of exhibits some form of collective organization, from the growth patterns of lichen to the complex electoral processes of modern democracies.

In a setting of a galaxy full of life, communication and transport technologies are the only barriers to the formation of galactic governmental organizations. Evidence based on ET activities on or near Earth indicating the widespread existence of such technologies suggests that galactic government is a very possible proposition. At The Farsight Institute we have often targeted an organization that we call the Galactic Federation. This is a target with many unusual aspects, and the breadth of stimuli associated with this target often demands much of a viewer's perceptual abilities. For obvious reasons, the headquarters of this organization is of especially great interest.

The target cue for this Type 3 session is written using an alternative cuing formula that was once used at The Farsight Institute. The exact specification of the cue is as follows.

> The Galactic Federation Headquarters Structure (at the moment of tasking): The viewer will clearly perceive and describe the structure housing the Galactic Federation.

The viewer will clearly sketch the structure housing the Galactic Federation. The viewer will clearly perceive and describe the location surrounding the Galactic Federation Headquarters structure. THE VIEWER WILL PERCEIVE ONLY THE INTENDED TARGET THAT IS CURRENTLY ASSOCIATED WITH THE ASSIGNED TARGET COORDINATES. THE VIEWER WILL NOT DESCRIBE ANY BEING, OBJECT, OR INTANGIBLE THAT DOES NOT EXIST IN THIS TARGET. THE VIEWER WILL REMAIN FREE FROM ALL NON-TARGET INFLUENCES.

16 December 1997
11:53 a.m.
Atlanta, Georgia
Protocols: Enhanced SRV, Type 3
Target coordinates: 1911/0325

I initiate my target contact with an ideogram that I perceive to be semi-soft, having both natural and artificial elements. I perceive the colors gray, black, and white. Some aspect of the target is light and compact, as well as irregularly shaped. I deduct the idea of a robot, and I sketch an oddly shaped object with many angles. There are two large squares within the object.

Probing my second ideogram, I perceive some aspect of the target to be hard and artificial. I declare this target aspect as a structure. Something is making a loud, sharp, grinding or rat-tat-tat sound. There are bright blue and white colors with high contrasts. My sketch is of a curved horizontal line with many rectangular structures below.

My next ideogram again feels hard and artificial. I sense the colors gray, tan, beige, and silver. There is a chirping sound. Some part of the target feels smooth, small, and compact. It is irregularly shaped. I deduct the idea of a space suit, and sketch another oddly shaped object, but this time with a convex, lens-shaped object inside.

Moving to Phase 2, I perceive sounds that I interpret as "bratting," or "rat-tat-tatting." There are rough textures with warm temperatures. Colors include blue, yellow, and green. The luminescence is bright with high contrasts. Somewhere there is a foul and pungent smell that reminds me of sewage. The magnitudes of the dimensions include short verticals, wide and long horizontals, as well as something sloping. Some aspect of the target is definitely curving or round. There is also the sense of something being open, as in open spaces, and I sense some type of energetics that feels like it is buzzing or humming. For my Phase 3 sketch I draw a curved horizontal line that resembles a planet's horizon as seen from space.

In Phase 4, I observe something rough and irregularly shaped. It is compact and small, and I draw the sketch of a curved structure within which are placed smaller rectangular structures. Inside the structure is something complex. I perceive many subjects, and I can sense their thoughts. These subjects are short, and I can perceive their clothes as well as their skin. Their heads are triangular-shaped. All of the subjects are of a similar type. I sketch the head and face of one such being, and deduct the idea of Greys. The face looks very similar to that of a typical Grey.

There is a great deal of subspace activity at the target site. The sense is that the activity is connected to the existing physical activity. Both physical and subspace activities are going on simultaneously with transparent coordination. There is a great deal of subspace technology at the target site. I deduct the idea of a plan. I again perceive many subjects, and I am aware of the floor of the modern structure where they are located. It is shiny and polished, and I focus my attention on a particular room. I sketch this room. There is one humanoid and many oddly shaped objects inside the room. There is a door and curved walls. It is a medium-size modern structure with technology inside. The room is very clean, and I perceive no dust. There is a great deal of subspace activity, and the subspace side of things seems more active than the physical side. The activity is not frantic, but it is very directed.

There is some kind of concept or idea associated with this target that characterizes the physical and subspace activity. I want to call this concept or idea "planning," but this is not quite right. It is as if the activity is already in progress, very active, not planning to be active. It is like a campaign. The emotions at the target site feel subdued, but with great or sharp interest. I again perceive small subjects that seem young, and I deduct "young Greys." These subjects not only feel very young and fresh, but also innocent in some way. There are many of them, and many more subspace beings than physical beings.

I am now perceiving huge numbers of subjects. They are

trying to initiate or trigger something. There is no sense of attacking to destroy. There is a concerted effort that is coordinated and determined.

There is a large group of subjects involved in a concerted campaign of highly directed intense activity. The sense is of an all-or-nothing intensity. Some type of total commitment has been made. There is no turning back. Retreat would be possible but undesired. I perceive the idea of "their right." There is an obligation at stake. It is the sense that the activity at the target site is owed. Someone has the right to collect on a debt. There is the sense of fighting for that which is morally and defensibly theirs.

I also perceive the emotions of anger and the concept of betrayal. The anger is collective. Something was done, a debt was incurred, and someone is trying to stop the collection on that debt. The collective anger is a result of the effort to steal that which is rightfully theirs. There is also some mild shock and indignation that is connected to the notion that the debt was incurred honestly and with innocence, in the sense that such a debt should be honored with equal honesty and innocence. There is anger specifically directed at someone or some group.

Again focusing my attention on the structure, I perceive many rooms with advanced technology within. I cue on the concept of activity and again perceive the idea of something that is directed or coordinated. Everything is organized. There is no play time here. I cue myself to move 500 feet up so as to change my perspective, and I sense that the concept "up" has no meaning. I then decide to move 1500 feet away from the target in an

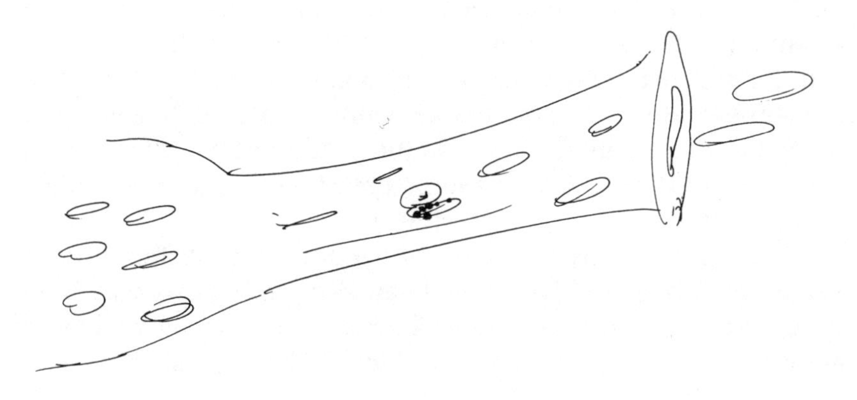

optimal direction to view the overall activity. I perceive structures that are modern and metallic. I sketch a large circular structure with many floors within. There are many subjects on the floors. It is a bright curved or round structure with many colored lights, lit up like a Christmas tree.

I then cue myself to move to the subspace activity that is at an optimal location for understanding the target activity. I sketch an oddly shaped object that resembles a flower vase turned on its side. There are elliptical or round objects inside this object. I deduct the idea of a wormhole. There is fast movement associated with transportation inside. I get the sense of being sucked through this object. It feels metallic and self-contained. I focus on one of the curved or elliptical objects inside the larger object, and I deduct the idea of a ship. This object seems isolated, not connected to anything physical. I perceive the color black. I then sketch this smaller object. The shape of the object is almost like that of a slice of pie. Around the object is not just a vacuum, it is nothing, almost like the thinness of time and space. The space seems black with some white. I cue on the purpose of the object and sense that its purpose is to fly.

Discussion

These session data are filled with descriptions of a large structure that has both physical and subspace aspects. The structure seems to be a nexus for activity and communication between distant places and groups. Currently, however, there appears to be activity within the headquarters that is very directed and intense. Some major campaign is underway. Apparently, the Galactic Federation Headquarters is not a place where beings sit around and amuse themselves with idle thoughts. It is a place of activity and struggle. It is a hub, an organizational center probably somewhat similar to our own United Nations. It seems that many Greys operate from within the headquarters. Interestingly, this headquarters also seems more based on the subspace side of existence than on the physical side.

What is meant by the sense of needing to collect on a debt? What was done that has since been betrayed? How was innocence answered with treachery? Does any of this have to do with activities relating to humans, and Earth? These questions are not answered by this session. Indeed, these data generate at least as many questions as they attempt to resolve.

Yet this much appears certain. The Galactic Federation Headquarters is a place of action. It is a meeting ground for activities that matter in important ways. Humans would be foolish to postpone knowing more about the activities of this organization. It may be that one day this organization will be summoned by ourselves to help us in our own struggle to survive and to grow as a species. Knowledge of the capabilities and the scope of this organization's influence may one day be crucial to our own future. Indeed, we do not understand the meaning behind the current activities at the headquarters. Perhaps some of the observed activity is associated with their efforts to assist us even now in our struggle to define our own future.

Chapter 25

BUDDHA

In my earlier book, *Cosmic Voyage,* I included a discussion of Buddha. My tasker had targeted Buddha, and a number of my remote-viewing experiences suggested that he currently has a leadership role in the Galactic Federation. As odd as this may seem at first, in retrospect, everyone has to do something after physical death. Moreover, it makes sense that an important personality such as Buddha would obtain a substantial role in a predominantly subspace organization that is attempting to influence events on Earth. My best guess is that Buddha has a leadership position that is involved in some way with the relations between Earth and the Galactic Federation.

The Type 3 session for this chapter came as a major surprise to me. At first I was not going to include it in this volume because I could not understand it. But eventually I realized that it fit with everything else I was discovering about current activities in our galactic neck of the woods. I now think this session is possibly very significant.

The essential clue and qualifiers for this chapter's target are as follows:

Buddha (current time). In addition to the relevant aspects of the general target as defined by the essential cue, the viewer perceives and describes the following target aspects:

- the target subject's state of mind
- the target subject's surroundings
- the target subject's activity

20 April 1998
12:02 p.m.
Atlanta, Georgia
Protocols: ESRV, Type 3
Target coordinates: 6978/5654

My first ideogram is of a structure that is curved and rounded, heavy and dense. The second and third ideograms are again of a dome or curved structure. I remember during the session being able to look up into the top of a dome shape that emitted bright light. I perceive a number of subjects below or within the dome. My fourth ideogram is of a subject who seems both middle-aged and heavyset. The final ideogram is of a man-made structure. My Phase 3 sketch is of a large globe near another curved surface. There is something pointed above the globe.

In Phase 4, I again perceive the large domed object. I also sense strong emotions associated with this target. There is land here with structures, and water is nearby. The intense emotionality of the subjects who are near the structures returns to my awareness. I perceive from the subspace realm a strong concern regarding the activities at the target site. The subspace side is rooting for something. There is fighting going on, a significant conflict. I perceive a cacophony of shouted voices. While I sense a conflict, I do not perceive any physical shooting, or the use of weapons. This does not seem like a military battle, but it is just as serious in its own way.

There is intense subspace activity associated with the events at the target. The subspace realm is deeply involved with this activity. There is a struggle here, and I deduct a war and a battle, although again I do not perceive that kind of fighting in front of me. But somehow this seems associated with a war. Many people are involved in this event. Subjects are moving around some structures. There are so many subjects that they cover all of the ground, including the spaces between the structures.

I move to the center of the target, and I perceive a male subject. I sense that this subject is important, and I deduct a king. This subject is a bit heavyset. He is inside a structure looking at something. It is dark inside the structure.

I execute a deep mind probe of the male subject, and I detect that he has strong emotions at the target time. He is upset with something. I deduct the Pope, a gathering at the Vatican, and Pope John Paul II.

I move 1,000 feet above the target, and I perceive land with structures. There are many subjects. There are loud noises, like shouting and horns. Lots of people are gathered between the structures.

I again shift my awareness to the central aspects of the target, and I perceive the male subject inside the structure again. The subject is experiencing intense emotions. Apparently he is upset.

Discussion

I remember the visual impressions of this session clearly. What troubled me at first about this session is that I could not understand why Buddha would be involved in a conflict that is possibly as severe as a war. Doesn't Buddha sit around and meditate all day? In retrospect, my understanding of the activities of evolved beings was certainly naive. If my interpretation of these data are correct, then I have a much better idea of how active all of our lives will be after we leave this physical existence. Heaven is not a gathering ground for the sleepyheaded. It is a place of action.

Apparently, Buddha and the Galactic Federation are deeply involved in an intense struggle that conveys the sense of a major conflict, perhaps a war. I do not know from the data in this session if the struggle is exactly the same as that associated with the renegade Reptilians, but I suspect the two conflicts are related. In my view, the most important aspect of this session is the very fact that Buddha is highly upset about the conflict that he is witnessing. This is a struggle that matters. It is not an idle squabble between children.

Whenever I turn my mind in the direction of the Galactic Federation, I perceive an organization in the grip of opposing tides of destiny. When I look at my own planet and perceive all of the conflicts that we humans have, I realize that the same must be true of those beings who live elsewhere. In an expanding universe, it would make sense that some of the various civilizations would end up fighting to control a larger share of the available resources, however these resources may be defined. Buddha is apparently involved in helping to resolve at least one of these conflicts.

PART V

The Martians

Chapter 26

Waiting on a Dead World

This chapter re-examines the issue of Martins still living on Mars. The target cue is designed to explore their living conditions, and their general state of security and well-being. However, one must always remember that when remote viewing, one's consciousness is exposed to all aspects that are relevant to an issue, and session results are often unpredictable as a consequence. With remote viewing, the subspace mind of the viewer often directs one's attention to aspects of a target that need to be included in a larger picture, which the specific target cue may only partially address. With this session we learn not just about the Martians and their current conditions, but about future events which could fatefully alter these conditions.

The essential cue and qualifiers for this chapter's target are as follows:

TARGET 3292/9537

Protocols used for this target: Enhanced SRV

The living physical subjects and their facilities that are currently located on Mars (at the time of tasking). In addition to the relevant aspects of the general target defined by the essential cue, the viewer perceives and describes the following target aspects:

- the physical environment of the subjects' living conditions
- the age and gender variations among the subjects

- the dominant groups among the subjects, including any governmental organizations
- the primary thoughts of the collective consciousness of the subjects
- the level of technology available to the subjects

19 May 1998
2:58 p.m.
Atlanta, Georgia
Protocols: Enhanced SRV, Type 2
Target coordinates: 3292/9537

I begin Phase 1 with the perception of two separate subjects. The first has something hard associated with it. The sketch of this subject suggests that the skin may have a hard or rough texture. The second subject seems quite human. The sketch of this subject suggests that he is a middle-aged or older male with a noticeable stomach. My next ideogram represents a structure, and I deduct the White House. The next two ideograms represent flags, and I deduct the American flag. In Phase 2, I perceive voices with electronic amplification. My Phase 3 sketch is of a complex houselike structure surrounded by a fence or wall. There are buildings to either side of the structure.

In Phase 4, I perceive land, a structure, and a fence. This is an important place, a center for government. I again deduct the White House. This is a complex structure containing technology that is communication related. The structure is near grass. From subspace, I perceive a sense of lively interest. It seems like the Galactic Federation, or something similar in subspace, has an observational window into or near this place. There is a high level of interest related to what is going on here. I have the guided deduction that whatever is going on may be related to the Reptilians.

Something on a large scale is happening. There is deviousness combined with conniving and dangerous trickery here. The wool is being pulled over someone's eyes. There is a plan, an agenda that is stark and sinister. I perceive a human subject, and I deduct President Clinton. (Remember that a deduction is a conclusion, not observed data.) There seems to be a presidential address in progress. (Which president is making the address is unclear.) There is an announcement about new friends, but this announcement contains deception.

I shift my awareness to the next most important aspect of the target. I perceive land that is flat and open. There are emotions here of panic and worry. This is a desert, dry and desolate, a barren world. I am now in an arid environment. There is dry sand everywhere, and it reminds me of the Lake Turkana region in East Africa.

I shift my awareness to a location that is optimal for understanding why I am viewing this scene. There are subjects and a structure in front of me. The structure is underground, and I am perceiving a hole. There is a poorly kept opening into the ground in front of me. It resembles a rundown mine shaft entrance.

I move into the opening. It is a hallway, a passageway, and there is a mixture of dirt and other materials on the wall. This place was patched together with very few resources. It feels desperate. Associated with the structure are the concepts of safety, haven, and refuge. It is a long, deep passageway.

Following the passageway I arrive at a room or large chamber. This is like a cave dwelling. There are unkempt subjects here. They seem to be wearing very worn garments. There is a woman, a child, a male, and I also perceive some type of light or fire. These subjects are living here at the bottom of this hole. They have no other place to go. It is like a subway system that is decrepit, and these are otherwise homeless people that live here.

Discussion

This session is best broken up into its basic components, of which there are three. First, from Phase 1, there are the physical images of a Reptilian and what appears to be a U.S. president. The second component appears later in the session in which the president seems to be making an address that contains false information. The session data suggest that the president has been tricked, and that there is a Reptilian influence associated with his behavior. Essentially, he is denying that something exists. The third component of this session is a group of subjects, indeed families, living in dire conditions underground. The surface of their world is barren and desertlike. This third component closely matches previously obtained remote-viewing data regarding the Martian dilemma.

My interpretation of these data suggest that the Reptilians are interested in maintaining the secrecy among humans about extraterrestrial life. This idea parallels the reasoning involved with their destroying the Mars96 Russian space probe. Apparently the Reptilians have an agenda that would be seriously disrupted should humans become widely aware of the extraterrestrial presence on this planet or on Mars. By remote-viewing this timeline, it is possible that this unhappy scenario of continued secrecy and disinformation will now be changed for the better.

Chapter 27

Go Tell It in the Mountain

Martian society, like all societies, is complex. Not all Martians are destitute. Some have access to technology and other resources, and others depend on the more fortunate. Some of the technology available to the elite seems to include ships capable of rapid interplanetary travel. Their travel schemes entail shuttling between Earth and Mars in order to transfer supplies. Once they arrive here, they need a place to stay, a facility for their Earth-based operations. Early remote-viewing results suggest that there is such a modern Martian facility in the United States located in the state of New Mexico underneath the mountain Santa Fe Baldy.

Remote viewers have targeted this facility many times, virtually always with the same results. I have personally grown to accept these results as basically accurate. One day, hopefully soon, we will all know the truth about these matters. No target is unverifiable forever. While we wait for confirmation of the existence of this underground facility, I suggest that you closely examine this Type 3 session and consider the possibilities yourself.

The cue for this target is written using an earlier formula for writing target cues, and the exact specification of the complete cue is as follows:

Martians under Mt. Santa Fe Baldy/Mt. Santa Fe Baldy, New Mexico (at the moment of tasking): The viewer will clearly perceive and describe the target subjects, only if they exist at the target location during the time period specified. The viewer will clearly perceive and describe the Santa Fe Baldy mountain in New Mexico. THE VIEWER WILL PERCEIVE ONLY THE INTENDED TARGET THAT IS CURRENTLY ASSOCIATED WITH THE ASSIGNED TARGET COORDINATES. THE VIEWER WILL NOT DESCRIBE ANY BEING, OBJECT, OR INTANGIBLE THAT DOES NOT EXIST IN THIS TARGET. THE VIEWER WILL REMAIN FREE FROM ALL NON-TARGET INFLUENCES.

18th December 1997
10:32 a.m.
Atlanta, Georgia
Protocols: Enhanced SRV, Type 3
Target coordinates: 1226/7191

My first ideogram takes me to a structure that is both hard and artificial. I perceive the colors brown and tan. There are aspects that are both light and dark at the target site. There is a grainy texture associated with the structure, and it is angular or squarish. My sketch of the structure resembles a three-sided square with a rectangle in the lower right resembling the door.

The second ideogram is also of a structure, this time feeling semi-hard and artificial. I perceive the color gray and feel textures that are smooth and polished. There are high contrasts in luminescence at the target, and something feels sharp. My sketch is of a rectangular object not dissimilar from a shoebox.

The third ideogram is of a long-haired subject. I perceive the colors light brown, beige, and dark brown. My next ideogram feels soft and artificial, but I cannot discern what it represents. I perceive the colors gray and yellow, and I sense some movement that is both smooth and silent. I sketch two convex curves that are opening outward with a narrow space between. There appears to be movement between the two curves. As a deduction I declare that there is a tubular squeezing effect here, sort of like squeezing a tube of toothpaste, but involving space and time. My final ideogram is again of a long-haired subject.

Moving to Phase 2, I hear voices, machinery, as well as grinding and cooking sounds. The textures are smooth and polished, and the temperatures are warm. The visuals include colors of gray, blue, yellow, and brown, with moderate levels of contrasts and luminescence. Tastes are salty, smells aromatic. The magnitudes of the physical dimensions are short and squat verticals, as well as wide and narrow horizontals, all variously combined. Something is curved, irregular, and angular. Something at the target feels heavy and dense, and I perceive energetics that are slow, pumping, pulsing, rhythmic, and undular. At this point in the session my feelings are those of curiosity. My Phase 3 sketch is of the rectangular structure with a long-haired subject wearing robe-like clothes.

Moving to Phase 4, I again perceive colors of gray and tan, and textures that are smooth and polished. There is something metallic here, and the site itself has some aspect that is large and open. I perceive land, a surface and the concept of something deep and underground. I sketch what appears to be rounded mountain with a bald top. I perceive rock, rocky textures, and I sense an associated magnitude of deep. There is an interesting perception of a wide spectrum of emotions at the target site. The emotions also feel very human. As a deduction I declare the mountain Santa Fe Baldy.

There are both males and females at the target site. They are wearing long white clothing, and I deduct lab uniforms. The subjects are walking about on a floor that has a patterned surface that is both polished and smooth.

There is some subspace activity at the target site. But this target feels more physical than subspace in focus. The subspace aspect feels like it is being coordinated with the physical in some fashion. But the physical is not recognizing, or perhaps it is not interested in,the subspace side. There is a subtle relationship between the two sides. I sense from the subspace side the concepts of overlord or protector, perhaps manipulator.

There is activity at this target location, and I deduct an ET base. I again perceive a smooth texture and the colors of gray and tan. The entire target is on a large-scale magnitude. Again I perceive a wide range of emotions. An angular structure is present, with rooms and hallways. I then draw a sketch that I label an internal schematic or layout of the structure. It has a circular circumference that feels much like a physical border or boundary. In the schematic is one large rectangular room, nine small, squarish rooms, and one circular chamber. All of these rooms and chambers are connected by passageways. The entire structure has the sense of a labyrinth that has many rooms, both large and small, in a cavernous arrangement.

Shifting my attention, I perceive one large opened chamber or room. I sketch the interior of the chamber. It has what appears to be an elevated walkway with a number of subjects standing on it. The walkway is a curved or domed structure that is somewhat elevated above the floor. To the lower right is a squarish structure.

There are large objects as well as many subjects in this chamber. It is a working environment, both modern and clean. Some of the objects are almost the size of trucks. Some objects are square or rectangular while others are curved. While observing the subjects that appear to be working in the room, I deduct lab coats.

The subjects feel very human. I note that their clothes really do resemble lab coats, in the sense of being both long and white. This is a sterile environment that conveys the concept of protection. Subspace activity that is both clandestine and supportive of the physical activity is going on at the target site. The room feels like a laboratory. I declare this both as a concept and as a deduction. This feels like a hybrid laboratory/hospital/operational facility. The facility is hidden and protected from obvious view, but it is known to others. I then deduct a secret military base of the type I have heard about in the deserts of Australia. As a guided deduction, I declare that this is an underground base.

I give myself a movement exercise that directs me to travel through the passageways in the manner that a camera would move while continually recording. There are hallways with polished surfaces. They are modern, squarish in structure, and clean. The rooms are not always apparent from within the hallway. One hallway leads to an intersection. There is an opening to a room connected to the intersection. I sketch this intersection and room. The room has little activity. There are many things on the floor, and I get the sense that the room is being used for storage.

I cue on the purpose of the facility, and I perceive it as an operations base. I also cue on the concept of ET activity, internally wondering if this is what I am perceiving. My initial impression is that the activity seems humanlike. Wondering why I was feeling something extraterrestrial about the target, I cue on the idea of ET objects or technology, and I perceive lots of this.

Trying to get a locational fix, I move 2,000 feet above the structure. It is cold and open. I perceive a clear blue sky. There is rough terrain below that is irregular, and I deduct Santa Fe Baldy and the idea of mountains. There is also the color green, and I deduct trees. Focusing further, I note that the terrain does appear to contain rock as well as trees and other foliage.

I then shift to the center of the target. I again find myself within a structure with rooms and chambers. There is work activity inside the structure, and the target feels both modern and advanced. I then move to the location that would be optimal to help me discern the primary purpose behind this choice of a target. I perceive and sketch a squarish box that is used for storage,

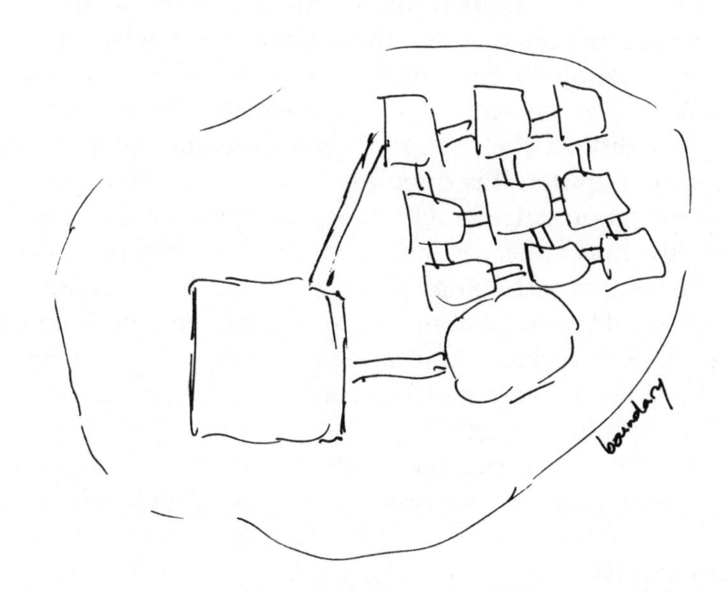

deducting a crate. There is technology here, and much is in storage. Much of this site involves something that is secret or hidden or perhaps not understood.

I then cue on the relationship between the subjects at the target site and the squarish object. It is the job of the target subjects to work with material that is found inside the object. I deduct that they are experimenting. I then cue on the relationship between the target subjects and any governmental link or connection. I perceive some type of governmental connection that feels Earthly and human. I deduct the U.S. government. But the governmental connection is separate in some way. It is very partial, incomplete.

Discussion

Remote viewers at The Farsight Institute (other than myself) have repeatedly noted that Martian ETs physically resemble humans. I have noticed that this resemblance is very strong, and thus it would be natural for me to perceive them as being human. The scenario of subjects resembling humans living in an underground facility and surrounded by ET technology that I observe in this session makes complete sense in the context of the Martian hypothesis.

There is no known open tunnel into the heart of this mountain. Thus, I must assume that those operating the base have an advanced technology comparable in some respects to that of other ETs who have been reported to be able to transform the physical quality of their ships at will, allowing them to pass through solid matter. It is often reported that UFOs appear and then disappear on radar, and that they are sometimes tracked flying directly into water or land at significant speeds. I do not know the level of technological development that is required to do this. But I do know that remote viewers have perceived what appears to be "cloaked" Martian ships as they fly directly into the rocky sides of Santa Fe Baldy, only to solidify and land in a large underground hangar.

The cue for this session has a time reference of the moment of tasking. Since it has been years since I have previously remote viewed this target location, it is interesting to note some of the changes that have taken place during this period. The first most significant change is the large number of target subjects walking around wearing what appear to be lab coats. I have never witnessed this before. The second major difference is the perception of some type of a link between the target subjects and a human government, potentially the U.S. government.

My current session leaves open the possibility that human ET contact has occurred. The clear human feel associated with this target in the current time, together with the observance of many subjects walking around in lab coats (which would be typical of Earth humans), would enhance the interpretation that humans are at some level currently interacting with the Martians based underneath Santa Fe Baldy. If this is true, I can only assume that it is a cooperative arrangement, however secretly maintained.

Chapter 28

CURRENT MARTIAN SOCIETY

This session focuses on Martian society itself. The cue is designed to locate the various political, spiritual, and dissenting groups in the society. Although Martian society may be small relative to the human population on Earth, it is nonetheless large enough to have significant internal variation among its groups.

The essential cue and qualifiers for this chapter's target are as follows:

TARGET 9028/3923

Protocols used for this target: SPP

Contemporary Martian Society currently living on Mars (at the time of tasking). In addition to the relevant aspects of the general target as defined by the essential cue, the viewer perceives and describes the following target aspects:

- the overall Martian society
- the most dominant political group in contemporary Martian society
- the most dominant spiritual group in contemporary Martian society
- the most primary dissension group in contemporary Martian society

15 May 1998
1:40 p.m.
Atlanta, Georgia

Protocols: SPP, Type 2
Target coordinates: 9028/3923

In Phase 1, I identify three groups, one large, one mid-sized, and a small elite group. All of the groups seem to be organized along some lines of authority, some loose and some rigid. In the Phase 2TM, I perceive that the target macro feels politically organized. It has a central leader who is a male. This leader seems to operate in a political fashion that is typical of how human democratic leaders interact with sub-macro groups. There is a high level of authority awareness that structures the hierarchy of the target. The idea of an authority guiding the larger society is deeply embedded in the populace.

Most of the activity of the populace is focused on the physical realm. There is a minimal use of telepathy for communication. The subspace realm is supportive of the physical realm; however, the physical side is somewhat confrontational with the subspace side. In Phase 3TM, I identify the three groups, and I label them in the conventional fashion, G1, G2, and G3.

G1

G1 has a small population with minimal physical variations within it. It is organized in an extremely rigid fashion. There is the sense that the group is very uniform or homogeneous. Some fear is connected with any deviations that may occur to shift the masses' allegiance away from the central authority.

A military flavor is associated with the consciousness of the significant leader for this group. This leader is a male, and he perceives himself to be a "good soldier."

I shift my awareness to a typical non-leader member of the group. Moving into this subject's mind, I perceive someone who follows orders, who stays out of trouble, and who holds emotions of worry, concern, and fear.

Negative emotions drive a significant amount of the behavior of this group. There is a great deal of anger and hostility, although I do not perceive toward whom the anger is directed. The group has an extreme interest in controlling something. The group members also remain hidden in some fashion from the remainder of the society. The macro-society does not know much about the operations of this group. On the subspace side, I perceive emotions of worry and concern. There is some plan or agenda on the subspace side relating to this group.

G2

The next group on which I focus in this society has a moderate population size. There are moderate levels of physical variations among the subjects, as well as some significant developmental variation in their subspace personalities. This group is one with a political orientation. There is a moderate level of concentration in authority, although the leadership of this group is somewhat limited in what it can force its members to do. This is a consequence of the degree of independence that the members of this group have.

The group seems to be involved with activities partially addressing subspace concerns, although this may be indirect. The physical side of this group's collective psychology does not appear to be deeply aware of purely subspace activities. In general, the group seems plagued with some degree of impotence, in the sense that the group does not seem capable of accomplishing things that are important to it.

The significant leader of this group is physically short, and a male. He is a lower-level leader. His status and authority are a result of the defaults associated with his position, and are generally not connected to his charisma or his personal leadership abilities. Shifting my awareness to a typical non-leader member of the group, I find a group participant with significant independence from the leadership, which this subject considers pompous.

G3

The final group that I examine in this session has a large population size with significant physical variations among its subjects. There is also substantial developmental variation among the subspace aspects of the subjects. There is a low concentration of authority, although it is uniformly distributed throughout this group. G3 is a loosely structured group with a high degree of group variation on the individual level in a variety of ways. The level of habitualized, institutionalized behavior among the group members is highly irregular.

This group is confused internally. There is a lack of control. The group is susceptible to opportunistic leaders who can emerge momentarily and cause a disturbance. The group's members are generally afraid and worried. Their psychology verges on the edge of panic.

The political tendency of this group is belief oriented. There is a high level of subspace activity among these group members, although their use of telepathy is low. This suggests that some of the subspace activity of this group may be religiously oriented.

I perceive a significant leader of this group, and I find him to be highly opportunistic. His position or status within this group is very uncertain, and it could change quickly. Shifting my awareness to a typical non-leader member of the group, I perceive emotions of fear, worry, and powerlessness. There is the sense of needing to "hunker down." In general, there is very little holding this group together other than a common experience or history. On the subspace side, I again perceive emotions of worry and concern as well as an intent to follow a plan or an agenda.

The Macro-Society Developmental Trajectory

I can identify three focal points in the history of this group. In the beginning, there was an orderly society with a clear political hierarchy and social structure. There was good cultural definition as well, in the sense that the multiple groups within the society functioned smoothly.

The end of the developmental trajectory is much different from the beginning. There have been a number of important collective experiences that have transformed this society significantly. Much of the society's development is based on these powerful experiences. The history of the society contains one highly significant dividing line.

Focusing on that point of transition, I perceive great structural change. It is the end of an older order. There was some crisis involving external forces that created an event. At this point there are interactions with subjects who are not members of the society. This interaction is a key aspect of their communal experiences. Consequent to this event, something new emerges, a new order, a completely new developmental direction.

Discussion

For a relatively small population, Martian society contains some interesting group dynamics. Of greatest interest seems to be G1, which may be a dissension group in the society. This group seems to be a highly organized group with a near military internal control structure. This does not necessarily imply that this group is aggressive, in the sense of wanting to attack others. But it certainly has a clear command and control orientation to its operations. The negative emotions associated with the members of this group may be reflective of the difficulties and frustrations that are experienced by many elements of Martian society.

Previously obtained data by both myself and other viewers suggest that the political/bureaucratic hierarchy of Martian society is stretched far beyond what might be considered normal operating limits. They have few resources, great needs, and only one option—moving their people to Earth, and then integrating them into our culture. It should come as no surprise that their nerves are frayed. Given this level of tension, it makes sense that some elements in the society would object strongly to the status quo, although they may not have any viable alternatives to the current situation.

The second group (G2) seems to be the dominant political group in Martian society. This group is apparently undisciplined, in the sense that the U.S. Congress is similarly undisciplined—individual members do not need to obey a higher authority, inasmuch as they have a significant degree of independence. In general, the group is not very effective with its activities. The activities are either not that important, on a day-to-day level, or they are not relevant to the primary concerns of the ordinary members of Martian society.

G3 seems clearly to be a spiritually dominated group, and I interpret these data to suggest that this particular group is the most reflective of the Martian masses. This large group is very loosely organized. They do not have deeply embedded ritualistic behavior that can tie all elements of the society together. The members of this group are vulnerable to the unpredictable whims of their government, including momentary appeals made by opportunistic leaders that may seem to spring up out of nowhere. Keeping such a group under control would indeed be difficult for any leadership.

This analysis of Martian society should come as no surprise given the dynamics of their situation. It would be helpful if humans could act to stabilize the Martian political dynamics by initiating some level of dialogue with the political elite. It is in the interest of humans to have Martian society remain as stable as possible in the near future. It is inevitable that all of these subjects will eventually need to be transferred to Earth. When that happens, we will need to work directly with these people. Their collective emotional state now will influence who we will be working with at that time, and it is always easier to interact with calm subjects who feel some level of support coming from the political establishment.

PART VI

Humanity's Choice of Destiny

OPTION #1

Chapter 29

The Reptilian Agenda

What do the Reptilians want from us? What do they want from Earth? In essence, what is their agenda here? Since I do not know their agenda, the best way to proceed is through an open-ended approach to a target cue. This chapter's session is based on a target cue that allows the subspace mind of the viewer to pick a target that will optimally reveal the Reptilian agenda for humans on Earth. This type of target cue strategy sometimes leads to very surprising results. For similar reasons, open-ended cues are also often a good thing to use when we think we know everything. The results using such cues sometimes let us see how little we actually understood, or how much we still need to know.

The essential cue and qualifiers for this chapter's target are as follows:

TARGET 2087/5903

Protocols used for this target: Enhanced SRV

The viewer perceives through the consciousness of the Galactic Federation Contact Person for The Farsight Institute, to remote view a target that will best represent the Reptilian ET agenda for humans currently living on Earth (at the time of tasking). In addition to the relevant aspects of the general target as defined by the essential cue, the viewer perceives and describes the following target aspects:

- the content of the agenda
- the underlying purpose of the agenda

17 May 1998
2:12 p.m.
Atlanta, Georgia
Protocols: Enhanced SRV, Type 2
Target coordinates: 2087/5903

As soon as I begin the session, I perceive a subspace subject surrounded by soft yellow light. Continuing in Phase 1, I perceive a different subject, a Reptilian. My next two ideograms represent a natural location with mountains, One shows a path in the mountains that leads to what appears to be an opening in the side of one mountain. In Phase 3, I sketch a mountainous terrain with a multi-level underground structure beneath one of the mountains.

Moving to Phase 4, I perceive foliage in an outdoor setting. I have the guided deduction of a base or facility. There is a path through the foliage. I appear to be on the surface of a land formation, perhaps a mountain. There is an open area that is sloping and cleared, as with a path. The dirt below seems like it has been stepped on, or driven on, repeatedly. The sun is bright, and the temperatures are warm. I do not perceive a physical subject here, but there are many nearby.

I execute a movement exercise to slowly slide to the location of the physical subjects. They are very deep underground. I am now in a structure with hallways. The walls are metallic and smooth. I note that this structure has many layers, or levels. It is dark here, but there are glowing lights. There are also subjects here, and they appear to be working. These subjects do not feel human, but they are humanoid. At least there is something odd about them. I deduct the concept of military.

I move to the center of the target and perceive many subjects in a closed area. There seems to be a lot of activity and interaction here. It feels like a college cafeteria in a way, lots of subjects doing things and making noise. It is a large chamber. There are horizontal things against the walls, like walking areas, elevated sidewalks, multiple layers.

I move to a location that is optimal for me to perceive the primary purpose for observing this scene. I am facing a subject who appears nonhuman. I deduct a Reptilian. I execute a deep mind probe on the subject. This person is a worker.

I move to the next most important aspect of this target, and I am again on the surface of the mountain. Moving back to the area with the many subjects, I again perceive rooms, passageways, and hallways. I cue on the concept of activity, and I perceive that this is a facility of some sort. Not much is going on right now.

I move to the time and location in this facility that will optimally reveal the primary activity that occurs here. I shift in time. This is a base in which subjects live and work long hours. It is difficult for them to go anywhere else. They just stay here and work.

I move to the location where the subjects are doing their work. There is sophisticated technology here. These subjects are simply working, and to them this is an ordinary workday. The light is dim here. I cue on the nature of the primary type of work at this facility, and I perceive test tubes, electronic equipment, laboratories, a scientific environment, and the concept of experimental research. This feels like a research lab.

I cue on the purpose of the research, and I perceive a young subject who is partially human, but still mostly of another species. I deduct Reptilian. I execute a deep mind probe on the partially human subject. This is a young kid. He has a lively intellect. His consciousness is very youthful, alive, refreshing, and awake. He has a healthy kind of emotionality.

I move to a typical subject in this facility who is not partially human, and I execute a deep mind probe on the subject. This being is mature and relatively slow intellectually. There is a certainness and quietness about the subject's consciousness, making it different from that of the partially human subject. There is a youthful brightness to the consciousness of the partially human subject, as well as a focus of awareness that seems to be different from, or missing from, the current subject.

I move around the facility and perceive hallways, chambers, and rooms. This is a very large place with lots of room, many subjects, and a great deal of activity. This place seems to have one primary reason to exist. I cue on the primary reason for this facility to exist, and I perceive that it is a place for governmental research operations. I do not perceive whose government this might be.

Discussion

This session is relatively straightforward. In my interpretation of these data, it appears that the agenda of the Reptilian extraterrestrials is to use the genetic stock of humanity to create a new race that is partially human and partially Reptilian. There is no indication in the data of this session to suggest what the Reptilians plan on doing with the remainder of humanity, if anything, once their new race is created. To know how humanity fits into their plans for the future (other than as a gene bank), we need to look into the future, which is what we do next.

Chapter 30

Life on Earth the Reptilian Way

The question left unanswered in the previous chapter addresses the future of humanity should the Reptilians succeed with their current agenda involving humans. Since I do not know what their plan for humanity is, it is again necessary to use an open-ended cue. For this chapter, the cue is designed to address various dimensions of the human condition on Earth should Reptilians have their way with this planet.

Recall that a great deal of remote-viewing data indicate that there is a complete continuum of realities and timelines, all of which equally exist. This is one of the reasons why it has historically been so difficult to remote view the future in particular. Many futures are nearly equally probable, and subtle and momentary considerations of a viewer's subspace personality can send a session off in any of these future directions. The possibilities are endless, and almost any scenario can be found in some future trajectory. For this reason, when we cue the future, it is necessary to state explicitly which timeline we want to be on in the session.

Since a future reality does exist in which humanity falls under a Reptilian influence, we can remote view that reality and discover the conditions under which those parts of our future selves are living. This is precisely what I have done in this session.

The essential cue and qualifiers for this chapter's target are as follows:

TARGET 6893/9086

Protocols used for this target: Enhanced SRV

The viewer perceives through the consciousness of the Galactic Federation Contact Person for The Farsight Institute, to remote view life on Earth in a future time frame in which the Reptilian ETs succeed with their current agenda involving humans (at a time in which it would be optimal to view the consequences to Earth of the Reptilian ET involvement with humans). In addition to the relevant aspects of the general target as defined by the essential cue, the viewer perceives and describes the following target aspects.

- the physical environment of Earth
- the condition of humans living on Earth
- the emotional state of humans living on Earth
- activities of humans living on Earth that may reveal the primary consequences of human and Reptilian ET interactions

6 May 1998
3:15 p.m.
Atlanta, Georgia
Protocols: Enhanced SRV, Type 2
Target coordinates: 6893/9086

I begin in this session by perceiving numerous subjects, each one at a time. Most of the subjects appear to be Reptilian, but one seems human with a military flavor. In Phase 3, I perceive the curved horizon of a planet with buildings and natural formations below.

In Phase 4, I perceive subjects, and emotions of pain. There is a great deal of activity in subspace. I perceive something circular in subspace, like a subspace planet. When I probe the physicals column, it feels shadowy and dim. When I probe the subspace column, it feels alive and filled with activity. In subspace I perceive subjects, land, a structure, and a circular surface. This feels like a subspace planet with structural features. There is bustling activity here. The subjects do not feel human. I perceive a subspace bureaucracy. This feels like a high point in the life of the subjects, or perhaps their civilization. There are subspace buildings and further construction. There are very new structures mixed with older structures. But there is a vibrancy here. This civilization is alive and well.

Moving to the most important center of government for this society, I perceive a large structure. I am near or on the surface of this place during the period of activity. There are lots of nonhuman subjects walking about.

I move to the primary subjects in the most important governing body while they are meeting, and I perceive both male and female subjects. There appears to be a matriarchal governmental situation. There is a large female nonhuman wearing some coverings or clothes. I deduct a Reptilian.

I execute a deep mind probe on the subject, and I perceive the concept of control. This Reptilian does not feel anger, but there is a certain sharpness. The mind does not merge well with mine. There is a cut-and-dried attitude with this subject. I perceive no softness to her personality.

I move to the next most important aspect of this target. I perceive an outside environment that is clean and healthy.

Again shifting my awareness, I move to the location that would be optimal to perceive the primary problems of the society. There are subjects in a room. They are experiencing pain associated with torture. There are severe human rights problems here. This society tortures subjects. This is a prison, and large numbers of subjects are incarcerated. This appears to be a way of emptying the society of problems. These prisoners have been locked up and forgotten.

I move to the next most important aspect of the target, and I perceive that I should shift my awareness to the target civilization and execute a collective deep mind probe. I do so. There are millions of subjects here, all struggling to survive. The consciousness of the subjects feels rough and harsh. But I perceive an innocence in the desire to understand existence, evolution, and life. These people are alone and besieged. They have been forgotten, abandoned, and rejected. They sense an intense anger and a need to fight. The subjects fight naturally. They appear difficult to work with. Give and take does not come easily for them.

Discussion

I do not understand all of these data. However, I suggest that they are best described in broad terms. Much of this session suggests subspace. This may be the result of a dimensionality shift in the time frame of the target. For example, these data most likely describe a future world that is derived from conditions very similar to those in which we live on Earth. Dimensional aspects of ourselves probably live, or have lived, on this world. But the world described in these data may reside along a timeline or in a dimension that is sufficiently distant from our own most probable evolutionary trajectory that it resides in a different vibrational state that I may have perceived as subspace. On the other hand, the suggestion of subspace may simply be a decoding error on my part that reflects the differentness of the setting described here from any other that I have perceived.

Regardless of the dimensionality question, this world has severe human rights problems. This is an unhappy world in which many subjects live under a government that exercises seemingly totalitarian control over its population. Incarceration is not foreign to millions of these beings. The closest analogous situation among humans that I know of would be South Africa during the height of the apartheid regime.

These data conform well to my previous observations regarding Reptilian society. The renegade Reptilians interacting with humans on Earth have an agenda that is not friendly to our existence. We humans have experienced totalitarian rule under our own hands. We know what that level of pain feels like. We have done it to ourselves; we do not need others to do this to us as well.

These data leave unanswered the question of whether the subjects I perceive in this setting are totally human or partially human. But does it really matter? Many of them are miserable, and we do not need that future for our children. It is worth fighting against. Over the span of humanity's existence, many have died to defend their families, friends, and nations against lesser threats.

Chapter 31

HUMAN SOCIETY ON A REPTILIAN TIMELINE

This chapter's session allows us to examine human society directly using the timeline in which the Reptilians succeed in their struggle to dominantly influence the future of humans and Earth. The target cue for this session is structured to identify groups that may exist on Earth in the future, but it is open-ended in terms of what those groups may be like. This session is perhaps the most revealing in understanding the implications for humanity of long-term interactions with the renegade Reptilian faction.

The essential cue and qualifiers for this chapter's target are as follows:

TARGET 4590/8945

Protocols used for this target: SPP

The viewer perceives through the consciousness of the Galactic Federation Contact Person for The Farsight Institute, to remote view human civilization on Earth in a future time frame in which the Reptilian ETs succeed with their current agenda involving humans (at a time in which it would be optimal to view the outcome of the Reptilian ET involvement with humans). In addition to the relevant aspects of the general target as defined by the essential cue, the viewer perceives and describes the following target aspects:

- the macro society that exists on Earth
- the human society that exists on Earth
- the Reptilian society that exists on Earth
- the most significant group that exists on Earth other than Reptilians and humans who are cooperating with the Reptilians

1 May 1998
2:15 p.m.
Atlanta, Georgia
Protocols: Enhanced SRV, Type 2
Target coordinates: 4590/8945

Phase 1 ideograms all represent groups that have political or administrative responsibilities. This is true of the macro as well as the sub-macro groups. The groups appear to be both large and small.

From Phase 2TM, the total population of the target society is very large. There are few physical and subspace variations among the populace, suggesting some homogeneity in racial types. There is a high concentration of authority that is sufficiently rigid to be labeled authoritarian. This society is a large organization or group that is highly divided. As an organization, there is considerable power or capability. But the organization has allowed itself to be split up into warring factions that have complete or near complete protection and autonomy from coercion. There is no "big picture" that governs the behavior of the subjects in the society. The operation of the macro-society seems to have degraded to a low point of extreme fragmentation.

There are multiple attractors within the society, which suggests that no single authority guides the populace with a vision. The common ideological theme of the larger society is clearly political, although its strength is weak.

Most of the activity of this society is focused on the physical realm. There is a very weak correspondence between the subspace and physical sides of existence. Communication across the subspace/physical divide is notably poor. The physical beings do not ignore subspace totally, but they pretend it does not matter. There is some disdain felt toward the entire subspace realm.

Sub-macro groups within the society are fighting. Competition between the groups is routine and expected. The groups fight for influence and survival. The macro-society is in a state of organizational atrophy. It has let the sub-macro groups divide up the larger society. The larger society has little control over the fighting among the sub-macro groups.

Focusing on a significant leader of the macro-society, I perceive a male physical subject. This subject is leading his own sub-macro group. He is fighting for his group against all other groups. Only a small fraction of his time is given to leading the macro-society. This subject seeks and needs to control the sub-macro groups, but he has little ability to do so. The subject is relatively impotent politically, and the sub-macro groups are insulated from him.

I shift my awareness to the collective consciousness of the macro-society. There is deep anger and resentment, as well as plotting and intrigue within this society. The target macro is filled with scheming sub-macro groups. It is like sitting on a pile of squirming worms. Emotions from the subspace realm include despair, isolation, impotence, and anger.

My Phase 3TM sketch identifies three groups, which I label in the conventional fashion.

G1

G1 is a small group, bureaucratically organized. Its membership and leadership have a very shallow perspective of any larger picture or vision of the society. This group seems numb in some emotional or psychological sense. Profound experiences register only barely on the psychology of this group's membership. The leadership also feels nearly numb psychologically.

The concentration of authority within this group is very high and rigidly enforced. There is only one attractor in the cultural topology of this group. This group's activities are almost entirely focused on the physical realm, and there is basically no functional relationship across the physical/subspace divide.

This group focuses on order, control, and manipulation of its membership. The larger macro-society is not aware of many of the activities of this group. The leader of this group is male, and there is a bureaucratic feel to his consciousness. This subject feels human or human-like. He holds a rigid collection of ideas that govern his behavior. His thoughts are not deep or penetrating to any significant degree. He leads because he is on the top of a chain of command. But he must build respect for himself from within his group.

Shifting my awareness to a typical non-leader member of the group, I perceive a male subject. He follows orders, and thus he himself is not technically responsible for his activity. He holds fear within his consciousness, and he tries to keep safe from authority and punishment. He is proud of his membership in G1, but he is also wary. He is aware that his group's activities could be questioned, and he is ready to leave his group should it be disbanded officially.

From a probe of the collective consciousness of G1, I perceive that the subspace realm is in a state of deep disharmony with the physical side. The subspace side seems to perceive reality differently than the physical side, and this leaves the subspace realm in a state of despair.

G2

G2 is a large organization or group with a relatively weak internal structure. The leadership is generally impotent politically. There is no mission that drives the psychology of this group. It is adrift, besieged, and weary.

The concentration of authority within this group is both high and authoritarian. The activities of this group are almost entirely focused on the physical realm. There is a serious breakdown in communication across the subspace/physical divide.

The significant leader of this group is a male. The subject is not capable of controlling the larger macro-society. He is the leader of this group by default, in the sense that there is no alternative to his leadership.

Shifting forward to a typical non-leader member of the group, I perceive fear, and the sense that the organization is out of control. Expanding my awareness to the collective consciousness of the group, I perceive the emotion of fear, and also a siege mentality. There is organizational impotence on the physical realm, in the sense that nothing significant can occur to change the broader picture.

G3

The final group on which I focus is a large sub-macro group with a political orientation. Members of the group have a relatively similar rank within the group. Movement within the group's hierarchy is slow, and each member has considerable autonomy. This group seems to have an inflated view of its relevance to the larger society—it is supposed to have an important leadership role, but for some reason it does not fulfill this role.

The concentration of authority within this group is high, and the quality of this authority is diverse. The group seems broken up into highly fragmented segments. Daily petty concerns drive the mentality of this group's membership. The activity of the group is almost entirely focused on the physical realm. The relationship between the physical and the subspace realms is broken, and nothing significant develops here.

This group interacts with the macro-society on the level of fraud, deceit, and trickery. A significant leader of this group is a physical male. Entering his mind, I perceive the concept of control. He is effectively organizing support on a small scale within this group. But he has an exaggerated view of his capabilities.

Focusing on a typical non-leader member of the group, I perceive someone who is "up-and-coming." He is almost on an equal level with the normal leadership. He is constantly challenging the leadership, in the way that bulls fight as they attempt to control the herd. Shifting my awareness to the collective consciousness of the group, I perceive fear and a shallow emotional depth. Profound emotional development is not common for this group.

The Macro-Society Developmental Trajectory

There are three significant points in the developmental trajectory of this target society. There is a founding point, a point of beginning. The organization of the society is established somewhat securely at this point. Near the end of the trajectory, there is a point of crisis. The organization or structure in the society survives, but there is a challenge to the original view. The time period between these two points is filled with irregularly spaced changes.

Not far from the founding point of this trajectory, there exists a point of division. The growth of the original society was stopped by someone at this point. The evolution of the society was then re-formulated and redirected. I deduct a civil war.

Discussion

This session is not difficult to interpret. Clearly in a future directed by the renegade Reptilians, human society on the Earth degenerates into tribal warfare. This is a world out of control, a world of disarray and disfunctionality. It makes no sense to attempt to find a silver lining to this future. This is as rotten as it can get, a dark age of chaos. Human against human, group against group, a world without a greater vision is a world that will die. While these data do not expand beyond the social chaos of that future moment, who can doubt the potential for the complete collapse of human civilization in this setting? If left to our own devices, perhaps humanity could rise again further into the future. Perhaps we could once again develop an inner vision of a species with a positive destiny. But who is to say that such a downtrodden society would be allowed to have a second chance? I can perceive no guarantee of our resurrection from such a dire world.

At this point in our evolution, we have a choice of futures. It is appropriate now to look at an alternative reality, a better—but still challenging—timeline that our current reality offers us.

OPTION #2

Chapter 32

THE GALACTIC FEDERATION AGENDA

It is now time to turn our attention to the Galactic Federation's agenda for humanity. The target cue for the current session parallels that which was used to explore the Reptilian agenda for humans as presented in Chapter 29. As before, the cue is open-ended, allowing the subspace mind to locate a target that best represents the Galactic Federation's agenda for humans living on Earth.

The essential cue and qualifiers for this chapter's target are as follows:

TARGET 2394/7902

Protocols used for this target: Enhanced SRV

A target that will best represent the Galactic Federation's agenda for humans currently living on Earth (at the time of tasking). In addition to the relevant aspects of the general target as defined by the essential cue, the viewer perceives and describes the following target aspects:

- the contents of the agenda
- the underlying purpose of the agenda

27 April 1998
3:11 p.m.

Atlanta, Georgia
Protocols: Enhanced SRV, Type 2
Target coordinates: 2394/7902

In Phase 1, I perceive a number of separate gestalts, including a pyramid structure, motion, a female humanoid subject with long hair, and a male subject who may be an ET. My Phase 3 sketch resembles a planetary horizon, pyramid structures on the surface of the planet, and motion above the surface of the planet.

In Phase 4, I perceive subjects and structures in an environment that looks desolate. I have the guided deduction of Mars. Some of the structures seem shaped like pyramids. They are heavy, nearly solid, with angles and smooth surfaces. Moving from structure to structure, I perceive one structure that is curved or rounded. It has a symbolic or representational purpose, almost religious but not quite. It is solid, and I deduct the Sphinx.

I move to the center of the target and perceive subjects who appear very human. They are in a group, and I perceive that they are wearing clothes. I move to a location that would be optimal for perceiving the central aspect of the target, and I perceive that the subjects are underground in some type of living environment. There are rooms and hallways here. Among the subjects, there are babies, females, and males. Given the age variation among the subjects, this seems like a living rather than working environment.

There appears to be a nursery and an area for communal living at this location. The subjects are experiencing hardship, and they are resigned to this situation. This place has tunnels and chambers that feel like they were designed with emergency survival needs at stake. This was not constructed as an optimal, desired, and well-thought-out place to live.

I move high up to the surface of the planet and perceive structures. There are facilities, warehouses, buildings holding technology and infrastructure. The subjects do not live on the surface, but the structures are connected to the activities of the subjects below. These surface structures seem to supply power to the subjects living below. The environment around the structures is generally wasteland and desertlike. There is very sparse vegetation, and some toxic waste. These facilities were rapidly constructed for the needs of the moment. I also perceive technology related to advanced communication, including radio.

I shift my awareness to the next most important aspect of the target, and I perceive advanced space technology in a modern scientific laboratory or operational facility. This technology is widely distributed in a large, enclosed area. There is the concept of communication with an outside agency or group. The subjects using this facility are signaling for help, or conducting negotiations. But the communications are somewhat one-sided. Clearly the subjects are reaching out for assistance. They are desperate. It is as if there is an acknowledgment of a serious mistake, and the subjects in the facility are assigned the job of "pulling a rabbit out of a hat." They are trying to fix something.

There is resentment and anger of some kind toward those who preceded them. But these subjects are resigned to the idea that this is their fate. They believe their efforts can change things. They have a set of beliefs or assumptions that keep them working. They have not given up.

I cue on the environment. It is seriously damaged or depleted. The surface of the planet is largely barren. There is something wrong with the atmosphere. I move along and around the surface of the planet. I perceive water with aquatic activity. There appear to be some structures on or near the water. The land is mostly dry and barren, although there is sporadic vegetation.

I shift my awareness to the next most important aspect of this target, and I perceive humanoid subjects. I execute a deep mind probe of the subjects. There is seething anger, and a sense of betrayal. The anger is overwhelming. I perceive the concept of treason.

Discussion

The beginning of this remote-viewing session contains data that suggest aspects of Mars, even the Cydonia complex that has been photographed on the Martian surface. But later sections of this session seem to be more clearly associated with Earth. It is possible that both Mars and Earth are being described in this session. It may be that Earth could go through a transformation that is not that dissimilar from much (but not all) of what occurred on

Mars in the distant past. My suspicion also is that the future of humans on Earth is somehow interrelated with that of the surviving Martian culture.

In my previous book on remote viewing, *Cosmic Voyage,* I present data that suggest that humanity may experience devastating environmental problems on Earth in the future. According to those data, the environmental problems will force surviving segments of our population to live underground in shelters, very much as the current surviving Martian culture exists to this day. My reading of the session for this chapter indicates that the Galactic Federation does not intend to intervene and modify this challenging future timeline. Apparently the Galactic Federation does not want to eradicate the problems that we bring on ourselves. Rather, the Federation is content to watch us struggle as we learn to control our own destiny.

The data in the current session contain no elements that suggest that the environmental devastation on Earth is caused by anyone other than ourselves. Moreover, nowhere in these data are there the concepts of human rights abuses, incarceration, or authoritarian rule. These are just humans struggling as best they can. They are angry at their predecessors for making selfish decisions that destroyed much of their world. But they are working to correct the problem, however severe it may be.

Such a future world is not Heaven on Earth. But it is our world, our Earth. It perhaps will be our destiny to create such hardships for ourselves, to survive such hardships however we can, and eventually to rise again. When we rise from such self-inflicted hardships, we will have learned a great deal about living, about growing as a species. We will emerge as a stronger and more mature race, a race capable of looking toward the stars to contribute to a growing universe. We will gain the wisdom of adversity. It is not our fate to live easy lives. It is our fate to grow.

My reading of these data suggest that the Galactic Federation will defend our right to evolve as a species, to make our own mistakes, and to learn from our own hardships. In essence, their agenda is to leave us alone, to let us find our own way in life. They will not care for us the way babies are coddled by a nursemaid, nor should we ever want that. They respect our freedom to learn, to grow, and to err. And I suspect they will be waiting for us with eager anticipation of our abilities to contribute to an expanding Galactic civilization when we once again rise off the surface of this planet, wiser, more loving, and with a deep inner desire to explore, and to serve our gradually maturing universe.

Chapter 33

Life on Earth the Galactic Federation Way

Now that I have a basic understanding of the agenda of the Galactic Federation with respect to humanity, which is basically to allow us to grow and mature with as little interference as possible from outside sources, it is time to ask how our lives will change as a consequence of an Earthly alliance with that galactic organization. Specifically, what will be the future consequences of direct human and Federation interactions? We can address this with an open-ended cue that allows the subspace mind to locate a target involving future life on Earth that will optimally reveal these consequences.

The essential cue and qualifiers for this chapter's target are as follows:

TARGET 4098/7543

Protocols used for this target: Enhanced SRV

Life on Earth in a future time frame in which the Galactic Federation succeeds with its current agenda involving humans (at a time in which it would be optimal to view the consequences to Earth of the Galactic Federation involvement with humans). In addition to the relevant aspects of the general target as defined by the essential cue, the viewer perceives and describes the following target aspects:

- the physical environment of Earth
- the condition of humans living on Earth

- the emotional state of humans living on Earth
- activities of humans living on Earth that may reveal the primary consequences of human and Galactic Federation interactions

7 May 1998
11:34 a.m.
Atlanta, Georgia
Protocols: Enhanced SRV, Type 2
Target coordinates: 4098/7543

In Phase 1, I perceive a male subject in what appears to be a metallic space suit. I also perceive a rectangular structure, and a female with long hair. My Phase 3 sketch is of a subject and what appears to be a space suit next to a cylinder.

Moving to Phase 4, I perceive subjects. The subjects are upset or disturbed in some way. Something is confined in a small area of a larger area. The subjects are wearing something bulky, like space suits. I perceive no land. It feels like open space. There is a tubular structure here, floating. I perceive that one of the subjects wearing bulky covering is a male. He is wearing clothes underneath this bulky covering.

I execute a deep mind probe on this male. He is working under time pressure. Something is at stake that is important.

I move to the location of that which is important for me to perceive about this scene, and I again perceive a male subject wearing bulky clothing working under duress. He has short hair. He is involved in some project to fix or repair something important, something connected to the concept of rescue.

Shifting my awareness to the next most important aspect of this target, I perceive a group of subjects in a meeting. They are on land in a structure. There is foliage nearby. This feels like a modern human city on twentieth-century Earth. The meeting is businesslike. The subjects are wearing modern suits. Their emotions convey seriousness and concern. The meeting is laced with the concepts of governments, regulation, and decisions. I deduct the U.S. president and the Congress. They are worried.

Cuing on the reason for their concern and worry, I perceive they are working on a plan to control something. They want to be involved. The scene is of a group of humans who are meeting to decide how certain things should be done. They are doing it because they have been brought into a discussion of important matters, and they are acting as their assigned responsibilities dictate they should act.

I move to the next most important aspect of this target, and I perceive a subject filled with anger. I deduct a Reptilian. The subject is nonhuman and male. I execute a deep mind probe on the subject and perceive that he is very assertive. He has access to great resources and technologies. He perceives these resources as a threat, or an ability to make a threat, against others.

This nonhuman subject seems to be semi-aware that I am viewing him. He is wondering if he is imagining this or if it is real. He is finding it hard to believe that I could perceive him. He concludes that I cannot get much and that the contact must only be superficial. This subject feels arrogant. He also feels Reptilian.

I shift my awareness to the location that is optimal for me to understand why I am perceiving this individual. I am alongside this individual. He is in a hallway in a structure. There is a room nearby with small subjects inside. They look like human children. They are at a table with small chairs. This reminds me of a day-care center.

There is another Reptilian sitting with the children, talking to them. I execute a deep mind probe of the Reptilian sitting with the children. This Reptilian is trying to learn about the human children through observation. The idea is not to control them, but to figure them out.

I cue on how these children got here. They were brought here by human authorities. I am now perceiving the subjects who brought the children, and they look very militaristic. They are wearing uniforms. This appears to be a day-care center for human orphans that is run by military authorities for the purpose of assisting the Reptilians in their efforts to interact and understand humans.

Moving to the next most important aspect of this target, I perceive a structure in space. I perceive what appears to be Earth nearby. I move into the structure. I do not have a bad feeling about this place. There is a consciousness here that is associated with this huge structure in space, and it feels good.

I move to the location that is optimal for me to perceive the reason why I am observing this scene. I perceive humanoid, but nonhuman, subjects. They are short and extraterrestrial. I feel no hostility here. I deduct the Greys. I get the feeling that the subjects here are totally aware of my presence. On a level of consciousness, I feel like a member of the crew, or a colleague.

There is a transparent physical/subspace relationship with what is going on here. The emotions feel deep, and I now clearly sense that this feels like a Grey place. I cue on why I am here. There is no assignment or purpose. I am supposed to make a comparison, and to observe. I sense that I am supposed to note the openness of the consciousness here. There is a complete lack of barriers. There is total honesty here, in the sense that the subjects are holding no secrets about themselves.

These beings want humans as friends, voluntarily. They want us to desire to be near them, and it should be that desire which creates a bonding between the two species that will last.

Now I am understanding the connection between this aspect of the target and the previous aspect involving the Reptilian and the day-care center. In the Reptilian/children scene, coercion was used against totally innocent and helpless children, orphans who have no parental protectors. Metaphorically, this is how the Reptilians want to interact with and "care for" humans. The Greys cannot operate on the level of coercion. They want us to interact with them with the innocence and openness of children, but as adults, and to choose to do so of our own accord.

Discussion

This is a complicated session that mixes a variety of themes. The material about space, rescue, and work under pressure probably represents human activity oriented toward space. It reflects our interest to work in space as well as our interest to interact with extraterrestrials. The material about the governmental meeting seems to address future political consultations involving the U.S. president and the Congress regarding decisions that need to be made involving the extraterrestrials. The scene with the Reptilian in the hallway suggests that the activities in the room with the military personnel, the other Reptilian, and the human children is a result of some coercion. Recall that the Reptilian in the hallway held the concept in his mind of threats related to resources which are available to him. I suspect that the

Reptilian interactions with human children in the observed setting were agreed to reluctantly by the human military authorities. They possibly feared (as a threat) the loss of technological assistance, perhaps weapons related.

The most telling aspect of this target is the comparison between the Reptilian interaction with human children in the daycare center and the Grey interaction with my own consciousness. The Reptilian mentality perceives humans as things to understand with buttons to push and knobs to twist. I perceive no sense that the Reptilians want to interact with us as equals. They want to understand us and to gain from us. But they do not seem concerned about our desire to determine our own fate, and to choose with whom we wish to interact. To me, the data in this session metaphorically represent an attitude that the Reptilians have toward us. We are children who need to be controlled. We need to be forced to do that which is useful or important to them, and that which they perceive as good for us.

I detect absolutely no sense of coercion, or any desire to force anything, from the consciousness of the Greys. Their open intellects are indeed childlike. But there is a deep maturity in their way of thinking that both respects our differentness and celebrates our common existence. I feel certain that if we were to reject the open invitation of the Greys to interact with us, they would not force the issue. But my observations suggest that the Reptilians will not leave without a fight.

Chapter 34

Human Society, on a Galactic Federation Timeline

It took me awhile to understand the data presented in this chapter. The first time that I looked at the session, I thought that there must be some mistake. Perhaps my tasker gave me the target coordinates for another cue, or perhaps I made some serious decoding errors, or perhaps there was just a mystery that I did not understand. My initial reaction was simply not to accept these data. I even put this session back on my shelf with a question mark on it. This just could not be our future. I would not accept it. Why would we voluntarily choose this future?

Upon reflection, I decided that I simply needed to continue to be totally honest with all of my data. I would do the best job I could to interpret these data, and then I would let my readers make up their own minds. But one thing is certain, our future is not what I initially expected before beginning this project, if these data are correct. My problem is that I could not easily let go of my idiosyncratic values. I could not let go of the past. I had difficulty accepting the fact that the future may be much different from my preconceived notion of our destiny.

The data in this session will speak for themselves. If I have interpreted them correctly, then I have let go of the past sufficiently to realize that our future will be guided by a vision greater than my own.

The essential cue and qualifiers for this chapter's target are as follows:

TARGET 3409/2390

Protocols used for this target: SPP

Human civilization on Earth in a future time frame in which the Galactic Federation succeeds with its current agenda involving humans (at a time in which it would be optimal to view the outcome of the Galactic Federation involvement with humans). In addition to the relevant aspects of the general target as defined by the essential cue, the viewer perceives and describes the following target aspects:

- the macro society that exists on Earth
- the human society that exists on Earth
- The most significant other ET society that exists on Earth

19 May 1998
11:22 a.m.
Atlanta, Georgia
Protocols: Enhanced SPP, Type 2
Target coordinates: 3409/2390

My first ideogram in Phase 1 suggests that the macro-society is large with many subdivisions. It is complex, but also homogeneous in some sense. I also perceive a number of sub-macro groups, varying in size.

Moving to Phase 2TM, I perceive a macro-society with some degree of homogeneity along at least one dimension, such as cultural, experiential, or racial. There are political divisions, but the intensity of the divisions is moderate to low. There is a focus on physical activity. The society feels "asleep" or quiet in some way, as if they are waiting, and perhaps have been waiting, for a long time.

The quality of authority in the society is diffuse and its concentration is weak. There are a few authority attractors that hold the attention of the society. The dominant ideology is political, and everyone is aware of this. But again, this ideological theme is weak in the society. The populace is divided into factions, and the members of these factions are bonded by mild group allegiance.

Most of the activities for the members of the society are focused on the physical realm, and language is used extensively. I perceive that the subspace realm is supportive of activities on the physical side, and there is intense interest among the physical beings in developing a subspace relationship. I perceive that the physical beings are a bit desperate in some fashion, and they may be looking toward the subspace side of existence for help.

The various factions within the macro-society participate with each other. They acknowledge differences in the society, and are generally supportive of group variation. The significant leader of the macro-society is a male. He is a bureaucrat or technocrat. He supports the sub-macro groups through his administration of social programs.

I perceive three groups in Phase 3TM. I label the groups conventionally.

G1

G1 is very small, and it may be a single subject. The subject feels masculine. Most major decisions for the society flow through him. He does not have a fully developed political orientation. He is more of a neutral but compassionate administrator. His activity is mostly in the physical realm. While he is not charismatic, he is respected for his administrative capabilities. He maintains a wide-angle perspective of the problems of his society.

G2

This group is small or moderate in size. It has a single or primary focus for authority, but that authority is not rigid. The group is generally homogeneous. The political orientation of the group is in its infancy. The activities of this group contain a significant focus on subspace. While they use language extensively, they also engage in telepathic communication. The group is admired by other members of the society. The group members are aware of their privileged status in the larger society, and they are somewhat protective of this status. Their minds are calm, but there is an underlying nervousness among them, deeply felt. G2 apparently has a mildly charismatic inspirational or religious leader.

G3

G3 is the largest of all the groups examined in this session. There are more numerous physical variations within this group, perhaps many of genetic origin. There is a diffuse concentration of authority, but the authority is strongly felt on the individual level.

The ideas that dominate this group reflect a focus on better times ahead, a salvation of some sort. There is a significant level of subspace activity with this group, and this activity may be spiritual in nature. The members of G3 hold all sorts of mixed-up ideas relating to subspace, many of which are simply superstitious. I sense that the subspace realm is very supportive of this group, even of their superstitions.

The other groups in the macro-society are a bit careful of G3. Currently, G3 is under control, but it could change and this group could become more volatile.

There is a male inspirational leader in this group. He is an opportunistic rabble-rouser with no specific agenda who is driven to his activities by boredom with his life. I get the sense that G3 is comparable to the lower class, or common masses. The common people feel primitive and fearful in some way.

The Macro-Society Developmental Trajectory

The beginning of the developmental trajectory is a peak period for the civilization. There are many subjects, religions, and even advanced artwork.

At some point after this peak period, there is an irreversible decline in the civilization. There is a time of great change.

Following this time of great change, there is a period in which the civilization hibernates or sleeps. This does not mean that the people in the civilization are hibernating. Rather, it is the culture or the civilization itself that is dormant.

After the period of dormancy, there is an awakening. The society emerges from the quiet period and begins new activity. During this time there is a brief period of development.

At this point the society leaves. They abandon their previous home. They simply pick up and go.

At the extreme end of the examined trajectory, there is abandonment. Ruins and relics of the past are all that remain in this place of origin. At the end of this trajectory, the previous civilization is essentially dead in terms of a linear continuation of the past. The society has left this place. They have begun to live in an entirely new way, and in a new location.

Discussion

What was so difficult for me to accept in these data is the development trajectory. The remainder of the session simply examines the society at a single point in its future. But the developmental trajectory is much more extensive in its perspective, and it is this wider view that contains the greatest surprise.

Earth is our current home world, our place of origin. It is also the nest that incubates humanity. But because of my remote-viewing experiences, I am beginning to view all of humanity as a hybrid between individuality and a collective. I sense that the individual nature of our personalities is an asset that will persist and develop through time. It is a positive ingredient to our overall sense of who we are. But we are also a large group with a collective consciousness. In a sense, we, as a collective, are somewhat like a single complex organism.

If one examines a hive of bees or an ant colony, it is easy to view these collective entities as a single organism. The individual bees or ants cannot exist outside the collective. Indeed, the various individual bees and ants can be viewed as organs in a larger entity, although the organs can walk about independently, and they are not all covered in one skin. But none of the parts can exist alone. These are extreme cases of collective entities.

It has often been mentioned that the Greys seem to exist as a collective. Their extensive telepathic capacities allow them to share one another's thoughts with no barriers separating individuals. In general, one might say that they are more collective than individual. Humans, on the other hand, seem to be more individual than collective, at least currently, and perhaps this will be a permanent feature of our aggregate personality. But both humans and Greys have individual and collective identities, and the differences between the groups are really variations in degree. Indeed, the current activities of the Greys suggest that they are attempting to incrementally move their collective consciousness in a direction that emphasizes the role of individuals.

What strikes me so strongly is that individuals form collectives, and the collectives evolve with a coherency that gives clear definition to the evolution of societies across time. There will always be individuals in any society that vary in their particular development, both physically in terms of age as well as with regard to the maturity of their personalities. But the societies are truly unique when compared across time. Humanity of the Dark Ages is totally different from humanity of the late twentieth century. When reading our own history as a planetary civilization, clearly the collective that we call humanity has evolved from a continuum of previous states to become something totally different today.

What this all means in terms of the remote-viewing session is that the collective entity that we call humanity is destined to leave this nest we call Earth. These data do not say that humanity will be defeated in any way. Rather, they indicate that our home world will one day have completed its purpose. We will not exist on this planet through all eternity. We will not forever depend on this rock to define ourselves. We will grow. We will hatch. And we will leave our protective shell behind.

In my reading of these data, I see a galactic existence for humanity. This is the true meaning of the Galactic Federation's agenda for us. We will not be the "Earth people" forever. We will be humans, and we will live throughout not only this galaxy but the universe. We have already seen this happening among other species. Apparently the Greys do not now live on only one planet. Their cities are ships in space. In the future, some of them may again locate themselves on a planet. But even that would only be for a distinct period of time. And what of the Martians? Are they not also planning to finally leave their world of origin behind? Can any species stay in one place forever?

It is probably true of all species that survive, that their collective destiny rests not on any one planet, but in a universe, and perhaps beyond. What was so hard for me to accept in these data is now obvious to me. We too will evolve collectively to a point where we must leave our place of origin. We too will abandon this planet one day. We too will live among the stars. This is our collective fate. As individuals, our destinies will ultimately be our personalties, and there will always be variations among us. But as a group, as a civilization, would we really want anything less than a destiny as cosmic explorers in a universe of challenge and wonder?

Chapter 35

WITH THE EYES OF GOD

For years now I have led a public battle to build an institute dedicated to developing our new science of spirituality. Some have taken our efforts seriously, but most have marginalized our work into the corners of their minds that they reserve for the bizarre and the ludicrous. Is it worth it to publicly confront a society with ideas for which they are so poorly prepared? Is it worth the public scorn, the professional ostracism, the financial risk, the personal pain?

People spend their entire lives running the race of time, living physically while pushing aside the question of the practical relevance of spirituality. Eventually they find themselves on their deathbed with the realization that all their achievements in the physical world are inconsequential relative to their ability to answer with certainty the question of whether their personalities will survive the demise of their bodies.

Readers, if you can hold in your mind the possibility that even one of the remote-viewing experiences that I have described in this volume actually happened, then you realize that remote viewing is not possible in the absence of a soul. Unless some aspect of ourselves is nonphysical, perceiving anything at a distance, or across time, would be impossible.

My experiences in consciousness have demonstrated to me beyond doubt that my soul is real, and that the soul of every human being on this planet is equally real. Since one does not physically "go" to any place when remote viewing, obviously there is something else about us that is already at the target location . . . at all target locations. What other mechanism could possibly explain the ability to perceive something accurately without any physical connection to that thing? Is there any possible way to explain the phenomenon of remote viewing other than to accept fully the scientifically verified reality of the unbounded nature of human consciousness, a consciousness that does not depend on the human body for its existence?

My remote-viewing experiences have also led me to understand God differently. I no longer look to some distant Heaven to find my Father. I now understand the meaning of the words once told by a wise sage: God is within.

I once had a remote-viewing experience where I perceived what I believe to be the essential beginning of the universe. There was a shape of light that somewhat resembled a spinning cloud or vortex. In the middle of the session, I questioned what to do next once I had perceived this enigma. In one of the columns of my Phase 4 matrix, I cued on the idea of a suggestion, inviting my subspace mind to offer a hint as to the meaning of this experience. The response was immediate. I sensed that I was to cue on God.

My conscious mind was confused by this suggestion, and probably as a consequence of this, the idea came to me that I should cue on someone who could explain this to me. I then cued on Jesus in the matrix. For reasons that I will explain, I have long considered him our older brother, in a literal sense. I suppose I intuitively felt that he would be sufficiently experienced and wise to know what to do in this situation. Instantly I sensed my awareness shift to include this being's flavor of consciousness, and I sensed him suggest to me that I should abandon fear and plunge into the vortex that I perceived. I did so.

The vortex was alive. There was the sense of a huge consciousness, and as I extended my mind across this consciousness, I felt stretched like the skin of a balloon, although not uncomfortably so. It then became clear to me that this being was terribly alone, and sad beyond measure. It had spent an eternity by itself, slowly evolving, until it finally grew to a point at which it could end its pain.

Then, in one sudden burst, I experienced this being's solution. The being essentially blew itself up, or at least much of itself. As I followed the outward rush of the being's fragmenting expansion, I perceived that it experienced a new joy that nearly overwhelmed me. The being did not die.

At first the bits and pieces of the larger being were too small and immature to even be aware of themselves. Neither were they aware of their own origin. From this point began the most profound evolution of the original being. It had become a parent to the fragments of itself.

The fragmented parts began to experience existence in a way that seemed independent of the parent. Initially they did not understand that they were literally part of a single larger being. Yet as they continued to grow in experience, they matured and developed an intense need to know how they came to exist, and indeed, the reason for their existence. This led them eventually to seek and discover the reality of their parent, their loving creator. It was at this moment of realization that they understood that they were their parent, and that their own growth and evolution was also the growth and evolution of their parent. The parent had created a way to look back at itself through a mirror of a multitude of individual consciousnesses.

OUR DIVINE HERITAGE

We have a divine heritage. What I have understood from my experiences in consciousness is that we are the fragments of God. By definition we are all made in his image. When I look at all those who surround me, I realize that I am looking into the face of my own Father. Now my readers can understand my motivation for pursuing research into consciousness. I view all human actions as part of God's own efforts to grow and to understand himself. Understanding consciousness is to understand God. I continue my research because nothing is more important to me than to know our Father.

This struggle to scientifically demonstrate our spiritual nature will ultimately succeed, but we need to differentiate individual battles from the war. What at first may appear as a total failure in one instance can later result in great achievement for the bigger picture. If the goal is to promote a global understanding of our composite nature as well as to demonstrate the fact of our divine heritage, then there can be no doubt of the eventual outcome. The spiritual urge of every soul to overtly recognize the truth will eventually drive all of humanity in the direction of self-realization. This would have to happen because it is in God's nature to evolve, and to eventually rediscover himself through the experiences of his children. To say that we will ultimately fail to see the truth about ourselves is to say that God will malfunction in his own evolution. In my view, this is not possible.

THE EXTRATERRESTRIALS

What role do the extraterrestrials play in this human drama of soul development? Ever since September 1993, when I had my initial exposure to remote viewing, I have become increasingly aware that many of those technologically advanced beings who are active on or near Earth have a deep respect for the human need to experience life without external coercion. I have been struck by the patience of these extraterrestrials as they gently prod a resistant human population that wants to bury its collective head in the sand like the proverbial ostrich.

In my opinion, if the Galactic Federation wanted to conquer this planet militarily, it could probably wipe out all human defensive infrastructure in a matter of minutes. The Greys alone, acting for the Federation, could do this without further assistance due to both their numbers (in terms of ships and personnel) and their extremely advanced technological capabilities. Yet my experiences with them suggest that the Galactic Federation is enormously concerned with our welfare, and they want to avoid any trauma that could negatively affect our future collective evolution. My perceptions indicate that they respect and value us and what we may become more than we understand our own potential destiny. However different these beings are from us, in my view, friends such as these are more valuable than anything I know.

In particular, from what I have been able to fathom of the essential nature of the Greys, they are driven by a desire that I have not seen matched elsewhere to evolve toward an ultimate communion with our Heavenly Father. To them, spirituality is all that matters. That, in essence, is what makes them so different from us.

The association of the Greys with the Galactic Federation makes sense to me. That larger organization accepts and even embraces the spiritual ideals of the Greys, or at least the right to pursue these ideals. But the Galactic Federation does not demand control over its participating species. For example, it did not prevent the Greys from ruining their own home world through wanton abuse of their environment. Nor did the Federation stop the Greys from originally interfering so dramatically with their own genetic makeup, an act from which they are still recovering today. From the perspective of the Galactic Federation, it is the right of a species to learn from its own mistakes, however severe these mistakes may be, as long as the species does not deprive another species of its own freedom to do likewise. The Federation does not want a universe of species that act like dogs on a leash. My observations indicate that the Galactic Federation embraces a philosophy that values the freedom of all species to achieve a unique destiny, to mature in their own ways through God's school of the living experience.

The Reptilians are a different matter entirely. The group of Reptilians currently interacting with humans appear to be a faction of a larger species, essentially a renegade group with a totalitarian political twist. I am completely aware that there may exist some people who have interacted with the Reptilians (including some UFO "abductees") who would describe their interactions as very positive. But my remote-viewing data suggest that the Reptilians as a group have an agenda for humanity that does not favor our interests as a species. They have their own interests in us, and they tend to strongly control that which they feel they need to control in order to serve those interests.

Perhaps the Reptilians view our penchant to destroy our environment as the irresponsible behavior of a species that could not survive on its own without a controlling and wiser guardian. They possibly view themselves analogously as prospective parents, and humans as orphans in need of adoption. Well, those prospective parents rule with a hard hand, and I for one would rather grow up without them. Should we fall under the Reptilian influence for an extended period of time, we would risk losing our ability to determine our own fate. We would trade our freedom for the caretakership that comes packaged with an enslaved race.

I do not think God would want to live the way the Reptilians want us to live. Because I view all of us as evolving fragments of our parent, I find it impossible to separate our own spirituality from the matter of the extraterrestrials. As a species, we are going to make a crucial decision in the near future. Our decision may be to accept the responsibility of our own destiny, which may include inflicting horrible environmental abuses to our world as we fight to understand ourselves and our true potential. Alternatively, we may abandon the difficult commitment to forge a future for ourselves that contributes meaning to a struggling universe, and instead choose a path that continues the pathology of controlling our masses with secrecy and manipulation.

OUR CHALLENGE, OUR DESTINY

There is no one person who will cause this planet to change. There is no one ET event that will cause all of humanity to go through that crisis of understanding. There is no one group that will force the major news outlets to report the truth about the extraterrestrials, the soul, or even God. For every person who makes an effort to awaken humanity, there are millions of others who are doing their part as well. Some of these others fly UFOs. Some are subspace beings who we label guardian angels. Others are common people who intuitively know that the truth has been hidden from them. Some petition Congress for hearings on these matters. Others write books and articles that call for greater openness.

As for myself, I will continue to speak. I will also continue my work at The Farsight Institute. The Institute will always present its research on its website, www.farsight.org. And I will ensure that the Institute maintains its commitment to an expanding research program from which so many have gained both knowledge and inspiration.

Eventually, our collective struggle will bear great fruit. Eventually, all of God's children will grow out of their state of confusion and accept the reality of a universe filled with life.

Eventually, all of our brothers and sisters will recognize our common destiny, and our divine heritage. When this happens, we will not be at the end of our evolutionary march. All the evidence suggests that we will then only begin to recognize the meaning of infinity, and thus of the truly long-term consequences to our Parent's gift of life to us.

Our struggle for self-realization is not the idiocy of a malfunctioning species. We face great challenges as humans because our Parent wants to grow with us, as us. It is not God's choice that we should suffer. It is his choice that He Himself should overcome all manner of obstacles. When we fight to assist humanity, we fight not only for the individuals that make up the collective, but for our Heavenly Father's desire to *be*. To the best of my understanding, there is no aspect of the universe that is truly separate from our Parent. Since we are made of Him, we will evolve as He evolves.

Why do I and others continue to fight a war of words against a tyranny of enforced ignorance? Could anything but a deeply internal love for our Father drive such behavior? For each of us, it is our choice to be like our Heavenly Father. We can see as He sees. We can hear as He hears. We can touch as He touches. We can love as He loves. Once one unravels the puzzle behind human existence, who would not fight to ensure a divine destiny as grand as ours?

Whenever we remote view, we extend our awareness across time and space. We can do this because we are already omnipresent. Remote viewing is an introduction to a greater future for our personalties. It proves that we have the ability to break out of our narrow intellectual confines and to see the universe more as God sees it. When we allow our minds to roam the heavens, we are truly, and literally, seeing with the eyes of God. When we learn from what we see, we grow both individually and collectively. With that growth comes a natural desire to evolve further, to see more, to learn more, to *be* more. This is the destiny of infinite positive evolution, which flows from our divine heritage.

We fight not to establish our right to explore and to grow within a wondrous universe. Rather, we fight ourselves to accept a right that was already granted to us. No one, but no one, can take this right from us unless we voluntarily surrender it. Let us seize this moment and define our evolutionary course for millennia to come. Listen not to the purveyors of fear. Fortune favors the bold. Let this be our mark in the universe. Let this be our destiny.

Communicating With the Author

Readers who would like to communicate with the author can send letters to the address below. Those who would like to know more about Scientific Remote Viewing can find additional information on The Farsight Institute's website, www.farsight.org. Information about the author can be found at his personal website, www.courtneybrown.com.

Courtney Brown
The Farsight Institute
P.O. Box 49243
Atlanta, GA 30359

e-mail: courtney@farsight.org

Appendix 1

Phase 5

Specialized procedures in SRV are performed in Phase 5. Below are thumbnail sketches of some of the Phase 5 procedures normally included at the end of the week-long introductory course.

Phase 5 requires a worksheet and a matrix, each on separate pieces of paper. The worksheet is labeled P5w, and the matrix is labeled P5m. The worksheet is positioned to the right of the matrix. All Phase 5 pages are assigned the same page number followed by the letters a, b, c, etc. for subsequent pages (such as 23a, 23b, etc.). The Phase 5 matrix is identical to the Phase 4 matrix. Also, P5 ½ matrix entries are made identically to P4 ½ entries.

1. **Timelines:** Have the viewer draw a horizontal line in the center of the worksheet. The viewer should then locate the target time, the current time, and the time of some significant event that is well known. The viewer should not be told the actual identification of the significant event, other than that it is event A. The viewer can also be instructed to probe the timeline for other significant events. Each event must be labeled generically, e.g., event A, B, C, and so on. The viewer should not probe for a specific year, only an event.

2. **Sketches:** Analytical sketches (more detailed than Phase 3 drawings) can be drawn and probed in the worksheet. Data obtained from the probes should be entered in the Phase 5 matrix. Lines can be drawn in the sketches to symbolically connect various places or objects. The viewer can switch from one place or object to another by alternately probing the separate parts of the drawing. Alternatively, the viewer can be instructed to move from one part of the drawing to another by following the line with his or her pen that connects the various parts. (See **sliding**.)

3. **Cuing:** In Phase 5, the monitor can suggest cues for the viewer to enter into the matrix that may be too leading for Phase 4. These cues can be from the viewer's Phase 4 data, or they can be the monitor's words. Again, cues originating verbatim from the viewer's data are entered into the Phase 5 matrix in parentheses ( ); data from the monitor in brackets [ ]. Moreover, all monitor-originating cues should have some obvious connection to the data obtained earlier so as to minimize the risk of "deduction peacocking," a phenomenon in which one deduction leads to another, and then another, etc., until a fictional storyline develops.

4. **Locational sketches:** The monitor instructs the viewer to draw a map, say, of the United States. No edge of the map should come within one inch of any edge of the Phase 5 worksheet paper. The monitor then says the name of a well-known location (usually a city). The viewer then automatically places his or her pen on that spot and quickly draws a line to the target location. No further monitor instructions are required other than to say the name of the original location. The line must be straight and rapidly executed. A slowly drawn or curved line indicates that the conscious mind interfered with the flow of the data.

5. **Symbolic sketches:** These sketches include some part or aspect of the target about which further information is needed. For example, using the Phase 5 worksheet, a circle can be used to represent a person being viewed, and a square can represent a governmental organization, and so on. The viewer is not told exactly what the symbols represent. Rather, the viewer is told a generic version of their nature (e.g., target subject, target group,

etc.). These generic identifiers are written near the symbols. A line is then drawn connecting the symbols. The line is labeled "relationship." Probes of the symbols (using the viewer's pen) and the relationship line yield information that is then entered into the Phase 5 matrix. If the symbols represent physical items, then the labels are placed in the physicals column of the matrix. The word "relationship" is entered in the concepts column in square brackets. All data are entered in the matrix.

Movement exercise for Phase 5: **Sliding:** The monitor can instruct the viewer to move from one location to another in a controlled fashion by having the viewer make a small circle on the Phase 5 worksheet. This circle should be labeled "A: location #1." Preferably, the viewer may write something more meaningful but still non-leading, such as "A: on top of the structure." Another small circle is then drawn on the worksheet in a position relative to the first circle such that this position is sensible.

For example, if the viewer is on top of a building, and the monitor wants the viewer to descend into the building, then the second circle would be below the first. The second circle is then labeled accordingly (e.g., "B: inside the structure"). The viewer is instructed to connect the first circle to the second circle with a line, and then to retrace this line slowly as needed in order to go back and forth between the two points. The viewer can also simply touch points A and B with his or her pen to shift quickly from one location to another. Alternatively, a cue placed in brackets (e.g., the words "building/inside") in the physicals column can achieve a similar result. However, sliding (down the line connecting points A and B) is useful if the monitor thinks that the viewer might profitably control the rate of movement, perhaps because the monitor suspects that observations made along the path of movement may be valuable.

Since there are no known distance limitations to this procedure, sliding is useful if the two locations are very far apart, such as two star systems. Often sliding can be used in combination with another technique. For example, the initial movement between two points can be accomplished with sliding, while subsequent movements can be quickly accomplished by having the viewer simply touch either of the connected circles. To enter data into the Phase 5 matrix, A and B are placed in the physicals column of the matrix inside square brackets, e.g., [A]. The data following A in the physicals column are related to point A in the Phase 5 worksheet. Data following B in the physicals column are related to point B in the Phase 5 worksheet.

Appendix 2

Enhanced SRV

The Farsight Protocols described in the previous chapters are called Basic SRV. In the advanced courses taught at the Institute, these procedures are modified significantly in order to exploit the greater capabilities that are possible with trained and competent viewers. These modifications are called Enhanced SRV.

Enhanced SRV resolves two problems inherent with Basic SRV. The first problem concerns an inadequacy in the use of Phase 1 data. Basic SRV collects and decodes a number of ideograms in Phase 1 that address various aspects of the target site. These ideograms are among the most important pieces of data in a remote-viewing session because the conscious mind has almost no chance to interfere with the collection of these data. Yet because the intent of the session is to proceed as quickly as possible to the later phases where more valuable data are collected, the Phase 1 ideograms are essentially thrown away as the viewer proceeds further into the session.

The second problem arises because viewers enter Phases 2 and 3 with a jumble of impressions left in their minds by all the gestalts in the various ideograms of Phase 1. For example, if four important target aspects are identified by four separate ideograms in Phase 1, from which aspect will viewers report, say, temperatures, and in what order? If the target is a campsite in

Alaska in the middle of winter, the viewer may report both the heat of the campfire as well as the cold of the surrounding snow. This mixture of gestalts continues throughout Phases 2 and 3, and viewers typically spend a great deal of time in Phase 4 sorting things out.

Enhanced SRV procedures resolve both of these problems. The enhancements also improve the quantity and quality of data that are collected throughout the session. They shorten the time needed to descriptively separate the various target aspects in Phase 4. The enhancements also produce operationally useful Phase 1 data relevant to each individual ideogram.

ENHANCED PHASES 1, 2, AND 3

Using Enhanced SRV, viewers begin their sessions by taking the target coordinates and drawing the ideogram in the normal fashion. They then write "A:" and describe the movements of the pen with words. The ideogram is then probed for primitive and advanced descriptors. Following this, the viewer writes "B:" and declares a low-level guess describing the gestalt that is reflected in the ideogram (such as "structure," "subject," "No-B," and so on). All of this is identical to Basic SRV.

The viewers then write "C:" underneath the B. The ideogram is then probed repeatedly, searching for low-level Phase 2 descriptors, but any data that is allowed in the Phase 4 matrix is also allowed here. Viewers do not force anything, allowing whatever is perceived to arrive freely. This method of probing is called "free response." You will remember that in Phase 2, data are always collected following a fixed structure (sounds, textures, temperatures, visuals, and so on). This fixed structure approach is still not used in Phase 1, but viewers can mentally remind themselves of a few of the categories of Phase 2 should they need assistance in initiating the flow of data. Probing the ideogram five or six times is often typical at this point, but viewers can probe the ideogram however many times as may seem appropriate should the data continue to flow. The data are entered vertically down the page.

As viewers collect more data under C, they will notice that a dim and vague mental image of the target aspect that is reflected in the ideogram begins to form. For example, if the ideogram reflects a structure, then the viewers will begin to develop an intuitive mental picture of the structure. Either directly underneath or to the left side of the column of data under C, the viewers then write "D:". A sketch is then made of this aspect of the target (such as a structure) underneath D.

All of the above is ideally done on one piece of paper. Thus, with Enhanced SRV, viewers obtain a complete collection of data for each ideogram, including a sketch. This solves the problem of having all of the ideogram-specific data being scrambled into only one Phase 2 and one Phase 3. But note that we have not yet "assembled the pieces."

Viewers then repeat the above process in normal Phase 1 fashion, taking the target coordinates between three and five times, seeing if any of the ideograms return subsequent to the appearance of a different ideogram. Most viewers tend to take the target coordinates five times since this allows them to obtain five complete collections of ideogram-related data, including five separate sketches. Once an ideogram reoccurs, or after taking the coordinates five times, viewers proceed to Phase 2.

Phase 2 is mechanically identical to that in Basic SRV, but now the viewer is free to "stand back" and look at the overall target site with a wide-angle perspective. The data are not limited to a particular gestalt (i.e., one ideogram). The sensory perceptions from all of the perceived gestalts compete (in a sense) for the attention of the viewer's subspace mind. Thus, the data that are perceived in Phase 2 are generally those that make the strongest impressions on the viewer's consciousness.

Phase 2 prepares the viewer to assemble the previously collected Phase 1 sketches into one composite sketch. This new sketch is performed in Phase 3. The Phase 3 page is positioned lengthwise (which, again, means the long side of the page is placed horizontally). Viewers can spend some time constructing their Phase 3 sketch, carefully contemplating the intuitive feel of the emerging sketch and placing each component in its appropriate place.

None of the previously sketched Phase 1 drawings need to be placed in the Phase 3 sketch. Indeed, many accurate Phase 3 sketches often do not appear elsewhere in the session. But viewers can place modified forms of any of the previously obtained sketches in the Phase 3 drawing should the intuitions be so directed.

ENHANCED PHASE 4

Enhanced Phase 4 is highly interactive and nonlinear. With Basic SRV, the structure is predominantly sequential and linear, taking the viewer from one step to another, allowing minimal structural flexibility. This limits the intrusion of the conscious mind into the data-collection process. Advanced practitioners of SRV are sufficiently familiar with both the structure of the session as well as the "feel" of the data such that they can take advantage of a greater degree of structural freedom as they interactively pursue their quest to understand the target.

Using Enhanced SRV, viewers work with five pieces of paper simultaneously. Each page is used to accomplish something different from that of the other pages. The first page is the normal Phase 4 matrix. The viewers work the matrix and go after the "Big Three" in the same fashion as with Basic SRV. However, there are some differences in the way viewers conduct other aspects of Phase 4, all of which are described below.

Tactile Probing

With Enhanced Phase 4, viewers extensively use their hands, and even their bodies, to explore the target. Once viewers have a mental image of the target, however fuzzy, they can then use their hands to "feel" the target, both externally and internally. With external probing, viewers tend to run their hands over the outline of shapes of things at the target site, like structures, mountains, and even faces. With internal probing, viewers press their hands (usually from top to bottom, although there is no rule here) through the target, perceiving internal aspects of structures, and so on. In one of my own sessions, I clearly perceived that a structure had three floors during an internal probe. I made this determination using my hands. I also perceived that there were subjects on the third and first floor of the structure.

Tactile probing is not limited to the use of the hands. One can also place one's head, or even one's entire body into the target at any given spot. For instance, in the example above, I then placed my head inside the structure to take a look at what was on each floor. This was done by literally bending my head forward while sitting at my desk and placing my head in the middle of the projected image of the structure. I then discerned that the top floor contained two subjects, one a male and the other a female. The bottom floor had a large number of subjects milling about.

Sometimes a viewer needs to explore a larger image of the target, or perhaps a component at the target site, such as a complex structure, or even a tunnel that goes through a mountain. To accomplish this, the viewer can back away from the desk and mentally project the image of the target into an empty area in the room. The viewer can then walk or crawl into the target or target component to perceive what is necessary.

After all tactile probing is completed, the viewer returns to the Phase 4 pages and enters the data in the appropriate places. If the data are verbally described, then the viewer enters the data as ordinary column entries, or as P4 ½ T entries. Here, the T represents "tactile."

Phase 4 Sketches

If at any time during the session a viewer obtains a visual image of the target, or an aspect of the target, the viewer must sketch this image immediately. Such mental images can arise during the process of probing the matrix, but they can also result from tactile probing of the target. In Enhanced Phase 4, there are three sketch pages. These pages are labeled Phase 4I, Phase 4E, and Phase 4L, where the I, E, and L represent "internal," "external," and "landscape," respectively. Instead of page numbers, the viewers write "a," "b," and "c," respectively, in the upper-right-hand corners. All pages are positioned lengthwise.

When perceiving a visual image, the viewer decides whether the image is internal or external. An internal image has a sense of being inside something else. For example, the viewer may perceive the inside of a room, or the inside of a piece of technology. If the image is the first obtained during Phase 4, the viewer places the letter A in the physicals column, and then circles the letter. The viewer then goes over to the P4I page, marks a corresponding circled A, and then draws the internal image.

If the mental image conveys the sense of being an external view, such as the outside of a structure, an object (say, a chair), a subject, or anything else, then the viewer follows the same procedure described above, but places the sketch on the P4E page. If this is the second sketch in Phase 4, then the viewer writes a circled B in the physicals column of the matrix, and on the P4E page. The drawing is then sketched near the circled B.

The Phase 4L page is similar to the Phase 3 page. Phase 4L is for putting pieces together. Many target aspects sketched on pages P4I and P4E can be located and redrawn in modified form in the P4L representation of the target. Phase 4L sketches are wide-angle representations of the target. The pieces can be assembled with considerable deliberation as well (that is, there is no reason to rush a P4L sketch). However, the viewer does not have to draw a detailed Phase 4L sketch. Nor do any of the P4I or P4E sketches have to be transferred to the P4L drawing. Sometimes a P4L drawing is simply a larger or more detailed version of the most important aspect of the target. But the goal is to create a P4L drawing that displays a more complete perspective of the target than is available in any other Phase 4 sketches.

The Phase 4 matrix and sketch pages should be placed in the proper arrangement before beginning Phase 4. All four pages are arranged in a rectangular pattern, like tiles on a kitchen floor. In clockwise order, the matrix page is placed at the lower left, then the P4I page, the P4E, and finally the P4L page, next to the matrix page.

This arrangement creates a fluid interactive working area. The viewer must not have to search for the correct page when the need comes to move to a particular sketching area, or when referring back to other aspects of the target.

Most viewers fill up multiple Phase 4 matrix pages. After the first matrix page is filled, that page is removed and a new matrix page is inserted in the same spot. If the page number for the first matrix page is 9, then the next matrix page is number 10, and so on. The sketch pages use letters. When the session is finished, all of the numbered pages are stacked sequentially first, followed by all of the sequentially arranged sketch pages.

When probing sketches (part of the "Big Three"), viewers sometimes use the back end of the pen rather than the point when probing is extensive. These data are often shown to others, and are sometimes displayed on the Internet as well as in print. In this way, advanced viewers avoid degrading the publication quality of their data by scattering too many probing marks on their drawings.

An Analytical Worksheet in Phase 4

It is often necessary to explore the target in Phase 4 using some of the analysis techniques of Phase 5. This is particularly true of symbolic diagrams that allow the viewer to describe relationships between various subjects, or between subjects and objects. Such abstract diagrams are not sketches, and thus cannot be placed on a sketch page. These are executed on a Phase 4 worksheet page, or Phase 4W page, where the W represents "worksheet." The viewer creates this worksheet together with the Phase 4 sketch pages.

The Phase 4W page is set lengthwise, and "P4W" is placed centered at the top of the page. The page "number" is d. The worksheet page does not need to be arranged in any particular place in front of the viewer. Normally it is kept to the side until needed.

A symbolic diagram in Phase 4 resembles that done for Phase 5. The viewer needs to draw two symbols (if there are two components to the symbolic diagram), label these symbols, and then draw a line between them and label this line "relationship." The viewer then enters the labels for each of the symbols in the Phase 4 matrix in the appropriate columns, all along the same horizontal row. The word "relationship" is placed in square brackets in the concepts column on the same horizontal row as the labels for the symbols. If one of the target aspects being explored is a subspace aspect, then the label for that aspect is entered in either square brackets or parentheses in the subspace column. The choice of square brackets or parentheses is determined by whether or not the word used to label the target aspect originates from the viewer's own data (which would normally be the case with a solo session). If both target aspects being explored are physical aspects (such as a subject and a structure), then the labels for both aspects are placed in the physicals column, separated by a slash, in one set of either square brackets or parentheses.

It is permissible to combine one square bracket with one

parenthesis if one label does not originate from the viewer's own data while the other label does. For example, entering "[central target subject/structure)" in the physicals column indicates that the words "central target subject" does not originate from the viewer's own previously obtained data, yet the word "structure" is an earlier matrix entry.

The viewer then probes the symbols on the P4W page, as well as the relationship line, and enters whatever data results from these probes in the Phase 4 matrix.

A New Level-Two Movement Exercise

Most target cues contain a variety of diverse qualifiers that address separate aspects of a target that the tasker wants explored. In order for advanced remote viewers to shift their awareness through these separate aspects, a modified form of a level-two movement exercise is used. The cue is as follows:

Move to the next most important aspect of the target and describe.

This cue is often used three or more times in a session. One stops using it when either repetition or tiredness appear. Advanced remote viewers do not use level-one movement exercises with as much frequency as novices, since they do not lose contact with the target as easily. Thus, advanced viewers have more time in the session to execute a larger number of level-two movement exercises. Experience has shown that the above level-two movement exercise is highly effective in assisting a viewer to obtain a wide variety of target data.

Binaries

Whenever viewers have a two-response question that needs to be answered in a session, they can use an advanced binary procedure to get the answer. To execute a binary, viewers put a letter (circled) in the concepts column of Phase 4, just the same way one would put a letter in the physicals column while making a sketch of something in Phase 4. Viewers then go to the Phase 4W page, write the letter (circled) and then do the binary procedure on that page. To do the procedure, viewers first write the question that needs to be answered. They then draw a long rectangle with a line down the center. The possible answers to the question are written at that time, one above each half of the rectangle. Viewers then put their pen in the center of the line that divides the rectangle, and the pen flies immediately to the correct side. An arrow head is added at the end of the quickly drawn line. Viewers then probe the centers of both halves of the rectangle to confirm their findings.

Binaries are very common in Enhanced SRV, especially near the end of a session. Some viewers even ask if they have satisfied the purpose of the target cue (or if they need to continue with the session). The following is an example of a binary procedure.

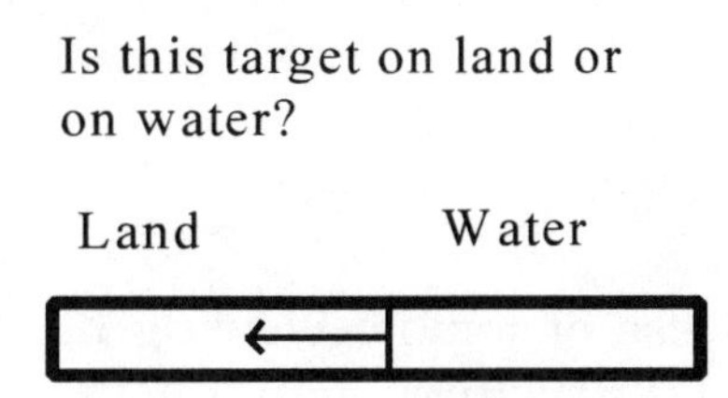

Appendix 3

Remote Viewing Social Systems

The traditional use of remote viewing has been to perceive physical objects, structures, individuals, groups of people, and activity. Only rarely has remote viewing attempted to explore societies or social organizations. The previously existing protocols were not designed to do this, and so even the best remote viewers were severely challenged in this regard. Since I am a social scientist by profession, naturally my interests led me to want to use remote viewing to examine politics and societies. I developed new remote-viewing protocols that directly address social and political concepts. These are called Social and Political SRV Protocols, or SPP.

This appendix explains how the new SPP protocols work. Readers should realize that our subspace minds are not limited to only describing physical information. Nothing is hidden from the human soul, even descriptive information regarding other societies. We not only can perceive places and events, we can use remote viewing to examine how entire societies operate regardless of their location, or even the time when they existed.

SPP Phase 1

SPP has five phases. It begins similarly to SRV, although preprinted templates are used throughout all phases of SPP. Copies of these templates can be found (free) on the Internet website, www.farsight.org.

Phase 1 of SPP is called "Macro Entry." The "macro" aspect of the target is its largest population unit. For example, if the target involves a country, the macro aspect would be the overall population of that country. The goal of Phase 1 is to describe the various groups that make up that overall population. Thus, we are "entering" the larger society and breaking it down into its sub-components, one component at a time. The choice of subcomponents is often determined by the target cue (which is not shown to the viewer until after the session is completed, of course).

In Phase 1, the target coordinates are taken as usual, followed by an ideogram. The ideogram is described in the normal fashion. The viewer then probes the ideogram and declares the basic descriptors, which are typically (1) physical or subspace, and (2) beings, subjects, or animals. With SPP we are fundamentally interested in describing the characteristics of organized living entities. It is therefore important to perceive what kind of entities we are examining, and whether or not the entities are physical beings living in physical space.

The viewer then probes the ideogram again in order to determine if the distributional characteristics of the target population are at the macro, sub-macro, or micro levels. If the ideogram is describing a population at the macro level, the viewer perceives that the information associated with the ideogram is of the highest level of social aggregation relevant to the target. For example, if the target was a multi-level approach to Israeli society, the macro perception would include the intuitive sense of the entire populace, both Jewish and Palestinian.

After probing the ideogram, if the viewer perceives a submacro quality, this indicates that the viewer is discerning distinctions between the separate groups in the society. Returning to the example of Israeli society, this could mean that the viewer is starting to perceive the separate Jewish and Palestinian populations within the Israeli society, or perhaps the sub-populations within these large groups. For example, among Jewish Israelis, there are a minimum of three distinct sub-populations (The Ashkanazi, Sephardim, and Falasha). Using another example, if the macro target was Belgian society, the two major sub-macro components of that society would be the Flemish- and French-speaking populations.

If the viewer perceives a micro-distributional characteristic to the ideogram, this would indicate that the ideogram represents the smallest aggregate unit within the population that is permissible given the target qualifiers. This could be small groups within a population, or perhaps even a single individual, although this is not the typical use of the distributional ideograms.

The viewer then attempts to identify the type of distribution that is captured by the ideogram. For example, the ideogram could describe the distribution of species within a population. On the other hand, the ideogram could identify the distribution of authority, culture, ideology, political orientation, or even group fragmentation of a society.

The viewer then probes the ideogram once again to perceive the distinct social components or groups that are associated with the target. If the viewer perceives descriptive aspects of these groups, then the viewer describes all of this in Phase 1.

In the final part of Phase 1 the viewer draws a schematic diagram of the society or the social component identified by the ideogram. The various parts of this symbolic diagram can be labeled in general terms.

Phase 1 is repeated between three and five times. With each repetition, the viewer takes the coordinates, draws an ideogram, and then probes the ideogram. Usually each repetition of Phase 1 addresses a separate aspect of the target population. Thus, for example, if the macro target was Belgium and Phase 1 is repeated three times, then the first pass may identify Belgian society, the second the French-speaking sector in Belgium, and the third the Flemish-speaking sector.

Phase 2TM

Phase 2TM obtains more detailed information of the largest unit (macro) of the target population. For this reason, this phase is labeled Phase 2TM (for target macro). If the purpose of the target is to describe the society of United States, then the target macro would be the overall population of the U.S.

The viewer enters information from top to bottom, typically probing on the punctuation (often a colon) at the end of each cue, as with Phase 2 in Basic SRV. But SPP uses some other probing techniques as well, such as focus ratios.

Focus ratios identify a binary division of a target populace. A focus ratio is the relative proportion of one type of activity when compared with another. For example, the subspace/physical activity focus ratio describes how much target activity resides in the subspace arena relative to the amount that resides in the physical realm. If the target was a prayer meeting, then one would expect the subspace/physical focus ratio to be higher (reflecting more subspace activity) than if the target was a football game (assuming, of course, that people are watching the game rather than praying for a victory).

Focus ratios can be used for many purposes. In SPP their primary usages are to estimate the level of subspace (relative to physical) activity and to identify the relative usage of telepathy for communication within a population when compared with physical language.

Phase 2TM also uses a specialized technique to analyze relationships. The first instance of this technique is in probing the collective relationship between the psychology of the subspace and physical aspects of the macro target group. The relationship procedure has three columns. The middle column is always labeled "relationship." When examining the subspace and physical psychological relationship, the left column is labeled "subspace" while the right is labeled "physical." The procedure begins by having the viewer probe the subspace column, and then draw an arrow to the center of the relationship column. The data that are perceived are entered into the relationship column. The viewer then probes the physicals column and draws an arrow from the physicals column into the center of the relationship column. The data, as before, are entered into the relationship column. This is repeated until the flow of data subsides.

Phase 2TM also uses this technique to explore the psychological relationship between the macro and sub-macro groups. In this case, the left column is labeled "sub-macro groups" while the right is labeled "macro-society." For example, if the macro-society was Germany during the period of the Nazis, the sub-macro groups might include Catholics, Protestants, Protestant peasants, Jews, etc. The relationship between the larger society and these groups would be perceived during column probes and subsequently entered as data in the relationship column.

The final specialized Phase 2TM procedure in need of description here is the "consciousness map." This is used to extract emotions and concepts associated with the collective consciousness of the target populace. This consciousness can have two aspects, subspace and non-subspace. "Non-subspace" is used as a label instead of "physical" since there is no need to assume a binary structure to all life. There may indeed be levels of existence within which many beings live that are not as dense or heavy as human physical reality, even though some such levels may be close to that of physical reality.

The consciousness map procedure uses both non-subspace and subspace columns separated by a circle with a dot in the middle. In each of the columns, there is a space for emotions and concepts. The viewer executes the consciousness map by probing the center of the circle (the dot) and then drawing a line to either the emotions or concepts space under each column. The circle represents the collective consciousness of the target populace. The dot in the center of the circle locates the viewer in the center of that collective consciousness (as compared with a peripheral location, say, within a sub-macro group).

Phase 3TM

Phase 3TM is a schematic representation of the target macro. Again, the "target macro" is the widest angle perspective of the target as it is defined in the target cue. The Phase 3TM incorporates both Phase 1 and Phase 2TM data. By the time the viewer completes Phase 2TM, the viewer is beginning to have a fairly complete perspective of the larger society as defined by the target cue, as well as many of the important groups that are located within the target macro. All of this is sketched in Phase 3TM.

Schematic representations of the target often employ a circle or other representative symbols, as well as lines that connect the symbols. The viewer labels each representative symbol. Each symbol typically represents a group within the target macro. The convention is to label the various groups in the target macro as G1, G2, and G3. It is not advisable to identify more than three groups at this stage, since a remote-viewing session using the

SPP protocols and three identified groups will likely take two hours to complete, which is about the maximum amount of time most people can productively spend remote viewing in one sitting. It is permissible to identify fewer than three groups.

It is often possible to understand how a society is organized by examining the schematic representation of the various groups within it. For example, if the schematic representation of the target includes a series of concentric circles, this would indicate that the society has a central core around which all other groups are organized. On the other hand, if this schematic representation includes separate circles, none of which have the same center, then the groups may be more autonomous in their organization, and there may not even be a central core to the target macro.

Phase 4GB

Phase 4GB follows Phase 3TM, and it closely parallels the structure of Phase 2TM. The "GB" in Phase 4GB stands for "group breakdown." After the target macro is sketched in Phase 3TM, the various groups that are identified in Phase 3TM are then examined sequentially in Phase 4GB, one at a time.

In the beginning of Phase 4GB, each particular group is identified. The identifying words are those that are used to identify each group's representative symbol in Phase 3TM. When Phase 4GB is completed for one group, a new set of pages are used to initiate the same data collection process for the next group, and so on.

Each repetition of Phase 4GB ends with a summary of the data in this section labeled "Phase 4GB OPEN." These summaries act as crucial points of synthesis for the viewer. The summaries allow the viewer to tie various points together that might otherwise be left unconnected given the sequential nature of the template.

Societies are not made up of isolated and separate individuals. Wherever there are sentient beings, they organize themselves. Groups and social structures are the natural outcome of subjects who interact with each other. These organized collectives have their own intelligence. Individuals participate in groups, and just as individuals make decisions, groups also make decisions. All of the subtleties of group intelligence are perceivable to the remote viewer. For example, the group intelligence of a riot is much different from that of a tea party. Similarly, the society of Germany under the Nazis during the 1930s was much different from Canadian society in the 1980s. The remote viewer typically perceives all of the component data for each of the various groups identified in each execution of Phase 4GB. In Phase 4GB OPEN all of the component parts can be brought together to more clearly describe the total sum of all of these parts.

Phase 5: Macro-Society Developmental Trajectory

Much remote-viewing evidence suggests that time does not exist. Rather, it appears to be a limitation of perception. When we live in the physical realm, we focus our perception sequentially, and events that are in the past are available to our minds only through memory. But when we remote view, we directly witness the actual events. Thus, all events in the past, present, and future still exist, and it is our perceptual limitations that create the illusion that only the present exists.

Phase 5 contains a line that, at first glance, appears to be a timeline. But time is irrelevant here. We are not interested in measuring months, years, or days. Rather, we want to describe the flow of history for a society. Phase 5 begins with the identification of a beginning and an end in a society's developmental history. These points are labeled A and Z. These points "bookend" the period of interest for the given society. The viewer then probes the line connecting points A and Z to determine the location of other significant points in the society's development. The viewer then enters the data for each one of these points in the appropriate spaces below the line. The viewer also describes the periods that lie in between the primary defining points. These periods are identified in Phase 5 by the two boundary points surrounded by square brackets (as with [AZ]).

Acknowledgments

Sandra Martin, President of Paraview, Inc., is my agent and friend. She is also a very beautiful soul, who has supported my efforts to report my remote-viewing research for the better part of a decade. She was with me when hardly anyone had heard of remote viewing. When times were rough she never doubted my ability to contribute something positive to the world. While she is a successful businesswoman, I have never known her to act selfishly. She lives for a higher purpose, driven by a desire to help our planet. Why I deserve to be represented by her, I do not know. But to those in the Heavens who sent her to me in this lifetime, God bless you.

I am grateful to the many people at Penguin Putnam who have supported my remote-viewing research. I have often wondered if it really could be luck that those who work with me are some of the best and most talented people in the publishing business. Some of those people have moved on in their careers. But I remember every one of them. In particular, I am grateful for the inspiration and assistance of Todd Keithly, Danielle Perez, Brian Tart, and Kari Paschall. There are many people at Penguin Putnam whose names I do not know, including line editors, publicists, and others whose work I have admired. I have been blessed to work with all of them.

Since 1996 I have been involved with an expanding group of

extremely talented remote viewers who have been trained at The Farsight Institute. By interacting with them, I have grown immeasurably, both as a viewer and as a person. I could not have written this book without their help and encouragement over the past few years. When I looked into their faces, I knew that all of this was real, and that all of our mutual struggles to share our experiences with a doubting world was worthwhile. In particular, I am grateful for the continued support of my friends and colleagues, Matthew Pfeiffer, Joey Jerome, Adele Lorraine, Richard Moore, and Denise Griffith. These are very spiritual people with deeply settled intellects. The many others whom I have not mentioned I also hold dearly in my heart.

Special thanks to my wife and my son. A wise African friend once told me that one's family is one's base. I did not understand this when he told me, but I have now experienced the truth of his words.

About the Author

Courtney Brown, Ph.D., is Director of The Farsight Institute, a nonprofit organization in Atlanta, Georgia. The Farsight Institute is dedicated to the development of the science of consciousness through remote viewing. Dr. Brown is the author of the first publicly available textbook in Scientific Remote Viewing®, a version of the remote viewing procedures that evolved from those developed in prestigious defense laboratories and used by the U.S. military for espionage purposes in the 1980s and 1990s. He is the author of five books, including *Cosmic Voyage: A Scientific Discovery of Extraterrestrials Visiting Earth* (Dutton 1996), which presents data collected using remote viewing to examine the extraterrestrial and UFO phenomena. Trained and still active as a social scientist, his work uniquely blends the pragmatic realm of governmental policy with the more esoteric matters of spiritual development and human interaction with extraterrestrial life.

· A NOTE ON THE TYPE ·

The typeface used in this book is a verson of Palatino, originally designed in 1950 bt Herman Zapf (b. 1918), one of the most prolific contemporary type designers, who has also created Melior and Optima. Palatino was first used to set the introduction of a book of Zapf's hand lettering, in an edition of eighty copies on Japan paper handbound by his wife, Gudrun von Hesse; the book sold out quickly and Zapf's name was made. (Remarkably, the lettering had actually been done when the self-taught calligrapher was only twenty-one.) Intended mainly for "display" (title pages, headings), Palatino owes its appearance both to calligraphy and the requirements of the cheap German paper at the time—perhaps why it is also one of the best-looking fonts on low-end computer printers. It was soon used to set text, however, causing Zapf to redraw its more elaborate letters.